T0255369

# PHILOSOPHY AND THE PRECAUTIONARY PRINCIPLE

Scholars in philosophy, law, economics, and other fields have widely debated how science, environmental precaution, and economic interests should be balanced in urgent contemporary problems such as climate change. One controversial focus of these discussions is the precautionary principle, according to which scientific uncertainty should not be a reason for delay in the face of serious threats to the environment or health. While the precautionary principle has been very influential, no generally accepted definition of it exists and critics charge that it is incoherent or hopelessly vague. This book presents and defends an interpretation of the precautionary principle from the perspective of philosophy of science, looking particularly at how it connects to decisions, scientific procedures, and evidence. Through careful analysis of numerous case studies, it shows how this interpretation leads to important insights on scientific uncertainty, intergenerational justice, and the relationship between values and policy-relevant science.

DANIEL STEEL is Associate Professor in the Department of Philosophy at Michigan State University. He is the author of *Across the Boundaries: Extrapolation in Biology and Social Science* (2008) and the co-editor (with Francesco Guala) of *The Philosophy of Social Science Reader* (2011).

# PHILOSOPHY AND THE PRECAUTIONARY PRINCIPLE

*Science, Evidence, and Environmental Policy*

DANIEL STEEL

*Michigan State University*

**CAMBRIDGE**
**UNIVERSITY PRESS**

University Printing House, Cambridge CB2 8BS, United Kingdom

One Liberty Plaza, 20th Floor, New York, NY 10006, USA

477 Williamstown Road, Port Melbourne, VIC 3207, Australia

314-321, 3rd Floor, Plot 3, Splendor Forum, Jasola District Centre, New Delhi - 110025, India

79 Anson Road, #06-04/06, Singapore 079906

Cambridge University Press is part of the University of Cambridge.

It furthers the University's mission by disseminating knowledge in the pursuit of education, learning and research at the highest international levels of excellence.

www.cambridge.org
Information on this title: www.cambridge.org/9781107435094

© Daniel Steel 2015

This publication is in copyright. Subject to statutory exception and to the provisions of relevant collective licensing agreements, no reproduction of any part may take place without the written permission of Cambridge University Press.

First published 2015
First paperback edition 2018

*A catalogue record for this publication is available from the British Library*

ISBN 978-1-107-07816-1 Hardback
ISBN 978-1-107-43509-4 Paperback

Cambridge University Press has no responsibility for the persistence or accuracy of URLs for external or third-party internet websites referred to in this publication, and does not guarantee that any content on such websites is, or will remain, accurate or appropriate.

献给国芳

*For Guofang*

# *Contents*


*List of figure and tables* *page* ix
*Preface* xi

1 The precaution controversy 1
1.1 Introduction 1
1.2 Putting the pieces together 9

2 Answering the dilemma objection 17
2.1 Introduction 17
2.2 The dilemma 19
2.3 A meta-decision rule 21
2.4 Proportionality 26
2.5 Two misunderstandings and an objection 37
2.6 Conclusions 42

3 The unity of the precautionary principle 44
3.1 *The* precautionary principle? 44
3.2 Arguments for disunity 46
3.3 Catastrophe, maximin, and minimax regret 49
3.4 Robust adaptive planning 62
3.5 Conclusions 68

4 The historical argument for precaution 69
4.1 Learning from history 69
4.2 The case for precaution 70
4.3 Risks and risks 81
4.4 An epistemic objection 89
4.5 Conclusions 94

5 Scientific uncertainty 95
5.1 Uncertainty about uncertainty 95
5.2 Defining scientific uncertainty 97

5.3 Probability and the limits of precaution 108
5.4 Conclusions 118

6 Counting the future 120
6.1 How to count the future? 120
6.2 Intergenerational impartiality and discounting 122
6.3 Arguments for the pure time preference 127
6.4 A new argument for intergenerational impartiality 135
6.5 Conclusions 143

7 Precautionary science and the value-free ideal 144
7.1 Science and values 144
7.2 The argument from inductive risk 146
7.3 Objections and replies 149
7.4 Rethinking values in science 160
7.5 Conclusions 170

8 Values, precaution, and uncertainty factors 171
8.1 Case study needed 171
8.2 Uncertainty factors 172
8.3 Replacing the value-free ideal 178
8.4 Epistemic precaution 192
8.5 Conclusions 197

9 Concluding case studies 199
9.1 Recapping the central themes 199
9.2 The precautionary principle and climate change mitigation 200
9.3 Recombinant bovine growth hormone 205
9.4 REACH 212
9.5 Future paths for precaution 217

*Appendix* 218
1 Formalizing the precautionary principle 218
2 Gardiner's Rawlsian maximin rule 225
3 Munthe's propositions 6 and 7 are equivalent 226
4 Triggering the precautionary principle 227
5 Chichilnisky and sustainable welfare 229

*References* 234
*Index* 254

# *Figure and tables*

## Figure

7.1 Cognitive utility theories of acceptance claim that, although our probability judgments affect both our decisions about practical matters and decisions about what to accept, what we accept should have no influence on practical decisions *page* 153

## Tables

3.1 Costs 57
3.2 Regrets 58
3.3 Qualitative rankings 60
3.4 Hansson's example 62
7.1 The possible states in Jeffrey's polio vaccine example 159

# *Preface*

In the course of working on this book I have sometimes been asked (especially by other philosophers), "What's the precautionary principle? And how did you get interested in *that*?" The short answer to the first question is that the precautionary principle is an influential yet hotly debated premise of an extensive body of environmental law, especially at the international level. Its central aim is to promote timely and reasonable responses to serious threats to health and the environment even in the face of substantial scientific uncertainty. Yet there is no generally agreed-upon interpretation of the precautionary principle, nor even agreement that a coherent and informative interpretation is possible. The second question might be read in a more general sense (i.e., why should a *philosopher*, and a *philosopher of science* no less, be interested in the precautionary principle) or a more autobiographical one (i.e., what got *you* interested in this topic). The autobiographical backstory traces to a number of discussions with a colleague, Karen Chou, from the Department of Animal Science at Michigan State University, about how to assess risks related to nanotechnology. I was interested in this topic as an outgrowth of my previous work on extrapolating scientific results from one context to another, such as from animal experiments to humans (Steel 2008). In the course of these discussions, I learned about something known as "uncertainty factors," a technical device for erring on the side of caution when estimating acceptable exposure levels to potentially hazardous substances. It seemed to me then – and it still does – that uncertainty factors are an example of the precautionary principle operating at an epistemic level. Moreover, the case was interesting because it was an established and widespread procedure in risk assessment rather than the pet idea of some small subgroup of scientists. Furthermore, coping with uncertainty is an unavoidable aspect of any effort to apply scientific knowledge to complex problems, such as environmental policy. So this set me thinking about a range of big and difficult questions. How should "scientific uncertainty" be defined? Just how should the precautionary principle be interpreted? How do epistemic manifestations of the principle connect to the topic, much debated in recent philosophy of science literature, of the role of values in science? I believe that the answers to these questions are very important to debates about how scientific knowledge should be brought to bear on a range of complex and pressing issues, such as climate change and processes for assessing the safety of chemicals and pharmaceuticals, to name a couple. And philosophy of science surely ought to have something helpful to say about such matters.

The literature on the precautionary principle (PP) is quite interdisciplinary. For instance, the legal scholar Cass Sunstein (2005) critically discusses the Rawls-inspired interpretation of PP proposed by the philosopher Stephen Gardiner (2006), and works by the philosopher Per Sandin (e.g., Sandin 1999) are widely cited in the literature on the precautionary principle within and without philosophy. Similarly, contributors to edited volumes on the precautionary principle hail from a variety of disciplinary backgrounds (see Fischer, Jones, and von Schomberg 2006; Raffensperger and Tickner 1999; Tickner 2003). This book leaps into this fray and engages with relevant literature regardless of the professional affiliations of its authors. As a result, the intended audience of this book includes anyone interested the interface of science and environmental policy in general and the precautionary principle and its implications in particular. Besides philosophers, these audiences include students and researchers in fields such as environmental economics, environmental law, and risk analysis among others.

Nevertheless, my own philosophical training and disciplinary association leaves a clear imprint on the book, both in terms of the issues raised and the lenses through which they are examined. Within philosophy, I expect the primary audiences for this book to be drawn from the ranks of philosophy of science and environmental philosophy. Among philosophers of science, the book should appeal to those interested in topics such as values in science and socially relevant philosophy of science. Interest in these topics appears to be growing, as suggested by a spate of recent philosophy of science books on policy-relevant science (Cranor 2011; Douglas 2009; Elliott 2011a; Kitcher 2011; Shrader-Frechette 2011). Interest in the precautionary principle among philosophers concerned with the intersection of science and environmental policies also appears to be rising (see Elliott 2010; Peterson 2006; Sprenger 2012; Steel 2013a; Steele 2006). For instance, a recent philosophy of science anthology includes a section on the precautionary principle (Bird and Ladyman 2012). And an article in a special issue of *Synthese* devoted to socially relevant philosophy of science lists the interpretation of the precautionary principle as a significant topic for this field (Tuana 2010, p. 481).

There is no sharp boundary dividing philosophy of science and environmental philosophy, as traditional philosophy of science topics are often deeply intertwined with environmental controversies. However, unlike philosophers of science, many if not most environmental philosophers approach their topic from a background of ethical theory. Nevertheless, this book is relevant to the interests of environmental philosophers of this sort. First, the precautionary principle is a recognized topic in environmental philosophy generally, as indicated by numerous publications on the topic (see Gardiner 2006; Hartzell-Nichols 2012; Manson 2002; McKinnon 2009; Munthe 2011; Sandin 1999). Second, since my approach to the precautionary principle integrates decision and epistemic aspects, it can appeal to philosophers whose primary interest lies in ethical dimensions of environmental issues. The book is written at a level that will make it useable as a text in upper-level undergraduate courses and graduate seminars. While engaging in depth with current philosophical discussions of PP and related topics (e.g., values in science), I generally avoid mathematical formalism in the main text and strive to explain central concepts in logical yet intuitively graspable ways. However, for those who wish to see the technical details, a formalized presentation of the interpretation of the precautionary principle advanced here, along with a few basic results, is provided in the Appendix.

Writing this book has been a long road, with many delays and unexpected turns. One pleasant consequence of this lengthy process is that there are many people for me to thank. My most significant debt of gratitude is unquestionably to my partner in life, Guofang Li, and our three children, Francis, Patrick, and Qiqi. The initial conception of the book predates my marriage to Guofang in June 2008 and our children's births. And its writing was so thoroughly enmeshed with these events and their aftermaths as to make it difficult for me to conceive the book separately from them. My best ideas can be unhesitatingly credited to the wonderful smiles and joy my family has given me. (I accept full responsibility for the mistakes!) And Guofang's love, support, and helpful advice on the ins and outs of academic publishing have been invaluable.

In addition, I would like to thank Kevin Elliott for reading and providing helpful comments and encouragement on drafts of most of the chapters. I am also grateful to Marion Hourdequin for serving as the commentator on a version of Chapter 2 presented at a Pacific American Philosophical Association session in 2013 and for stimulating philosophical discussion over Thai food afterward. This interaction helped sharpen ideas in Chapter 2 and inspired some of the discussion in Chapter 4. Thanks are due as well to Derek Turner for a commentary in *Ethics, Policy and Environment* (Turner 2013) on a paper (Steel 2013a) that is the backbone of Chapter 2. I am grateful to Kristie Dotson for reading and providing encouraging feedback on a draft of Chapter 6. I thank Paul Thompson for help with the bovine growth hormone example in Chapter 9. I thank Kyle Whyte for his collaboration on an article (Steel and Whyte 2012) whose ideas fed into Chapter 8. I am grateful to students in a graduate seminar taught at Michigan State University in 2011 on topics relevant to this book for their helpful feedback and to Kevin Elliott and Heather Douglas for speaking at Michigan State in connection with this seminar. I would also like to thank Michigan State University for an Intramural Research Program grant (IRGP-1448) awarded in 2008 that supported the early stages of this project. I would also like to thank audiences in various venues and locations who have heard and responded to presentations of work linked to this project. These include the Philosophy of Social Science Roundtable in 2009, the Department of History and Philosophy of Science at the University of Pittsburgh in 2011, the Three Rivers Philosophy Conference at the University of South Carolina in the same year, the Great Plains Society for the Study of Argumentation Conference at Iowa State University in 2012, the Biennial Meeting of the Philosophy of Science Association in 2012, the workshop on Cognitive Attitudes and Values in Science at the University of Notre Dame in 2012, the Pacific APA in 2013, and the Institute for Resources Environment and Sustainability at the University of British Columbia in 2014.

Finally, express my thanks to everyone at Cambridge University Press who helped this book become a reality. I am grateful to the two anonymous reviewers of the manuscript, especially the initially skeptical one who pushed me to express the central ideas of the book more perspicuously and to develop some arguments with a greater degree of rigor. Thanks also to Hilary Gaskin for prodding me to think of an improved title.

The work in this book is largely previously unpublished. Six of the nine chapters – specifically, Chapters 1, 3, 4, 5, 6, and 9 – appear here in their entirety for the first time. Nevertheless, the other chapters draw upon already published material, although this work is usually reorganized, rearranged, and often rewritten. Chapter 2 and much of discussion in ensuing chapters develops ideas from Steel (2013a), Chapter 7 reuses ideas and material from Steel (2010) and Steel (2013b), and Chapter 8 borrows from Steel (2011) and Steel and Whyte (2012). The copyright holders of these articles are as follows:

Copyright © 2010 by the Philosophy of Science Association. This article was first published in *Philosophy of Science*, 77 (January 2010), 14–34.

Copyright © 2011 Elsevier. This article was first published in *Studies in History and Philosophy of Biological and Biomedical Sciences*, 42 (2011), 356–64.

Copyright © 2012 The Johns Hopkins University Press. This article was first published in *Kennedy Institute of Ethics Journal*, 22.2 (2012), 163–82. Reprinted with permission by Johns Hopkins University Press.

Copyright © 2013 Taylor & Francis. This article was first published in *Ethics, Policy and Environment*, 16 (2013), 318–37.

Copyright © 2013 by the Philosophy of Science Association. This article was first published in *Philosophy of Science*, 80 (December 2013), pp. 818–28.

I thank the publishers for permission, or the right granted in the publication agreement, to reprint portions of these articles here.

CHAPTER 1

# *The precaution controversy*

> In order to protect the environment, the precautionary approach shall be widely applied by States according to their capabilities. Where there are threats of serious or irreversible damage, lack of full scientific certainty shall not be used as a reason for postponing cost-effective measures to prevent environmental degradation.
>
> *Principle 15 of the 1992 Rio Declaration on Environment and Development*[1]

> When an activity raises threats of harm to human health or the environment, precautionary measures should be taken even if some cause-and-effect relationships are not fully established scientifically.
>
> *Wingspread Statement on the Precautionary Principle*[2]

> The precautionary principle may well be the most innovative, pervasive, and significant new concept in environmental policy over the past quarter century. It may also be the most reckless, arbitrary, and ill-advised.
>
> Gary Marchant and Kenneth Mossman[3]

## 1.1 Introduction

Since the 1980s, the precautionary principle (PP) has become an increasingly prevalent fixture of international environmental agreements, from chlorofluorocarbons to biodiversity to climate change.[4] But despite – or perhaps because of – its prominence, PP is also extremely controversial. While PP is the subject of a massive academic literature, it remains notoriously difficult to define and responses to basic objections remain unclear.

[1] See the United Nations Framework Convention on Climate Change (article 3.3) for a very similar statement.

[2] See Raffensberger and Tickner (1999, pp. 353–4).

[3] See Marchant and Mossman (2004, p. 1).

[4] See (Raffensberger and Tickner 1999; Fischer, Jones, and von Schomberg 2006; Foster 2011; Trouwborst 2006; Whiteside 2006).

This book presents and defends an interpretation of PP from the perspective of philosophy of science. As a philosopher, I am concerned with basic problems of interpretation, logical coherence, and rationale. And as a philosopher of science, I am particularly concerned with how PP connects to scientific fields, such as climate science and toxicology, whose research comes in close contact with controversial environmental issues. Although PP has been approached from a variety of perspectives, I believe that philosophy generally and philosophy of science specifically has something valuable and important to offer. Objections to PP typically boil down to issues of a deeply philosophical nature, a number of which pertain to basic questions concerning values and scientific research. Can PP be formulated so it is both a distinctive and rational approach to environmental policy issues? What is its basic rationale in comparison to other approaches? How does it differ from approaches with which it is often contrasted, such as cost–benefit analysis? How does PP interact with policy relevant science? And can it do so in a way that does not threaten the integrity and reliability of scientific research? This book develops an interpretation of PP that aims to provide more adequate answers to these questions.

Much valuable work has already been devoted to achieving a better understanding of PP. A number of authors have examined the logical structure of the principle (Manson 2002; Sandin 1999), dissecting it into conditions concerning harm, knowledge, and a proposed remedy. Various perspectives or possible types of interpretation of PP have been distinguished (Ahteensuu and Sandin 2012; Sandin 2006). For instance, PP might be construed as a meta-rule that imposes general constraints on how decisions about environmental policy are made, as a decision rule that selects among concrete policy options, or as an epistemic rule requiring that a high standard of evidence be satisfied before a new technology is accepted as safe. Others have explored the relationship between PP and related concepts, such as the maximin rule (Ackerman 2008a; Gardiner 2006, 2010, 2011; Hansson 1997), robust adaptive planning (Doyen and Pereau 2009; Johnson 2012; Mitchell 2009; Sprenger 2012), and alternatives assessment (O'Brien 2000). In addition, much work has been done on explicating the role of PP in international law (Foster 2011; Trouwborst 2006).

Nevertheless, I believe that some fundamental challenges have not been adequately addressed in previous literature on PP. I discuss three of these now to provide a foretaste of the line of argument that this book will pursue: (1) the lack of an adequate response to the objection that, depending on how it is interpreted, PP is either vacuous or irrational; (2) absence of an explanation of how the many ideas and perspectives associated with PP fit together as a coherent whole; and (3) an adequate account of the relation between PP and policy-relevant scientific research. Let us consider these issues in turn.

Critics of PP often charge that it can be given either a weak interpretation – according to which uncertainty does not justify inaction in the face of serious threats – or a strong interpretation – according to which precaution is required in the face of any scientifically plausible and serious environmental hazard. Weak interpretations are said to be true but trivial, since no reasonable person would demand complete certainty as a requisite for taking precautions. On the other hand, strong interpretations are claimed to be incoherent, and hence irrational, because environmental regulations themselves come with some risk of harmful effects and hence PP often precludes the very steps it recommends. I refer to this argument against PP as the dilemma objection.

Most responses to the dilemma objection focus on the second horn. One common reply is to propose that PP should qualified by a *de minimis* condition, which specifies a fixed evidential threshold that must be crossed before the principle is triggered (see Peterson 2002; Sandin 2005; Sandin *et al.* 2002, pp. 291–2). However, this reply is inadequate, because incoherence still arises whenever the harmful effects of the precaution themselves attain the evidential standard set out in the *de minimis* condition. In such circumstances, PP would recommend both for and against the precaution, the *de minimis* condition notwithstanding.

A number of other responses to the incoherence horn of the dilemma objection can be found in the literature. For example, Per Sandin (2006, pp. 179–80) suggests that applications of PP must be understood in relation to a context in which a particular type of danger is salient. To illustrate this idea, Sandin considers the practice of prescribing antibiotics as a precautionary measure for patients undergoing surgery. In this context, prescribing the antibiotic is precautionary only with respect to a possible infection, and not with respect to possible harms of excessive antibiotic use such as the evolution of resistant strains of bacteria. Thus, Sandin argues, PP does not generate contradictory recommendations if it is understood in its proper context. However, this is a problematic response, because there is no discernible justification for making a decision in a context wherein plausible and significant harmful effects of a proposed action are disregarded. In Sandin's example, it is entirely reasonable to insist that antibiotic resistance be relevant to decisions concerning best practices for prescribing antibiotics. To claim that PP is applied in contexts in which negative effects of the recommended precaution are "off screen" seems only to reinforce the position of critics who assert that the incoherence of PP is often overlooked because of the regrettable human tendency to fixate on a single threat at a time (see Sunstein 2001, chapter 2; Sunstein 2005, chapters 2, 3, 4).

Sandin's "context" response might be read as suggesting that the incoherence objection is only an example of the general problem of local versus global framing of decisions (see Sandin *et al.* 2002, p. 293). For example, the decision above could be framed locally as whether to prescribe antibiotics as a prophylactic *for this particular patient* rather than more globally as whether such prescriptions should be made *generally* to patients in similar circumstances, and a decision rule might lead to opposite results in the two cases. If this were all the incoherence objection amounted to, then defenders of PP could easily reply that the problem is not specific to PP but confronts all decision rules. However, there is no need to interpret the incoherence objection in this manner. The objection is most naturally construed as charging that PP can lead to incoherent results *within a single framing of a decision problem.* For instance, critics would assert that PP leads to incoherent results when applied to questions about best practices of antibiotic prescription, recommending both for and against.

Another line of response to the incoherence horn of the dilemma turns on a deontological distinction between positive and negative duties (John 2007, p. 222; Weckert and Moor 2006, p. 199). A positive duty is an obligation to do good things, while a negative duty is an obligation to refrain from doing harm. It is commonly thought that negative duties are weightier than positive duties: for instance, that it is worse to murder a person than to fail to rescue her. However, this response is also problematic because PP is primarily intended to justify regulations, for example, that restrict the use of a toxic chemical. But regulations cannot be justified by appeal to negative duties, because enacting a regulation is not an omission but instead an action implemented by a government agency (see Munthe 2011, p. 71). The obligation to avoid harms resulting from regulations, then, would be a negative duty, and the argument in favor of a regulation would be grounded in a positive duty of the government to protect the environment or public health.

Another response is that the incoherence horn of the dilemma fails due to not noticing the role of proportionality in applications of PP (see Fischer, Jones, and von Schomberg 2006; Whiteside 2006). Proportionality recommends that "precautionary responses ought to correspond to the perceived dimensions of the risks involved" (Trouwborst 2006, p. 150). Hence, "ban it!" is far from the only policy option available to PP, as some versions of the second horn of the dilemma seem to presume. I think that the concept of proportionality is in fact the key to turning back the second horn of the dilemma. However, this response is not adequate without some further elaboration. Granted, outright bans are not the only type of policy PP can recommend. But that alone is not an adequate answer, because the second horn of the dilemma does not require the assumption that precautions always take the form of absolute prohibitions. Instead, it turns on the possibility that the proposed precaution has potentially harmful effects that would be sufficient to trigger an application of PP, which would in turn recommend that the precaution itself be avoided or substantially restricted. Whether or not proportionality effectively addresses this issue is unclear given the rather vague terms in which it has been formulated. So the concept stands in need of further development if it is to serve as an adequate answer to the charge of incoherence.

I turn now to a second major challenge confronting interpretations of PP. This challenge is the great multiplicity of sometimes apparently conflicting ideas associated with it. To take just one example, consider the maximin rule, which several authors have suggested as a basis for interpreting PP (Ackerman 2008a; Gardiner 2006; Hansson 1997). The maximin rule recommends that one select the policy option that has the least bad worst-case outcome. However, other authors have suggested that PP should be understood in relation to the concept of minimax regret, according to which one should choose the action that minimizes the maximum shortfall from the best that could have been achieved (Chisholm and Clarke 1993). Yet minimax regret and the maximin rule can easily lead to conflicting results (see Hansson 1997). If PP is construed as a principle that aims to generate useful policy guidance, this simply will not do. It cannot be identical to a pair of contradictory principles. Of course, the same point holds for any other conflicting ideas associated with PP.

Some advocates of PP do not regard the absence of a unified account of PP as problematic. For instance, Lauren Hartzell-Nichols (2012, p. 160; 2013) proposes that, rather than one precautionary principle, there are many, each designed for a distinct set of circumstances. However, I do not think this is a stable position. For what makes all of these different precautionary principles instances of the same general type? On the one hand, if a substantive answer can be provided to this question, then it seems that some unification of PP is afoot after all. On the other, if no substantive answer can be provided, then "PP" would be little more than an empty label that can be applied to almost anything one likes. But such a situation would render calls to adhere to PP found in international environmental agreements vacuous. Another disunified approach suggests that, despite its name, PP is in fact not a principle at all but rather a "repository" in which to deposit "adventurous" ideas that challenge conventional approaches to environmental policy (Jordan and O'Riordan 1999, p. 16). But such an approach faces obvious difficulties. For how do we decide which ideas may be dropped into the precautionary grab bag? And what should we do when those ideas conflict with one another? Answering such questions would require articulating some general conception of what PP does and does not assert, which the "repository" approach to PP explicitly disavows. But without answering such questions, the "repository" approach merely lends support to critics who charge that PP is no more than empty rhetoric masquerading as a serious approach to environmental issues.[5]

Some advocates of PP have attempted general interpretations that encompass a wide range of approaches. Arie Trouwborst (2006) attempts to distill the central elements of PP on the basis of an extensive review of formulations of the principle found in international law. This effort is, I think, extremely valuable insofar as providing a sense of what is generally meant by PP in a wide range of international agreements on environmental issues. As such, it imposes some general constraints on what a philosophical interpretation of PP should look like. In Trouwborst's account, international environmental law treats PP as a genuine principle – contrary to the "repository" approach described in the previous paragraph – involving several components, such as proportionality and the "tripod" of a knowledge condition, harm condition, and recommended precaution. But these general outlines are not sufficiently specific to resolve either concerns about the multitude of potentially conflicting ideas associated with PP or the dilemma objection. For instance, questions about the relationship of maximin and minimax regret in regard to PP are not answered, and while Trouwborst describes the incoherence horn of the dilemma objection, no answer to it is proposed (2006, pp. 184–7).

[5] Indeed, Marchant and Mossman (2004) cite Jordan and O'Riordan (1999) to support just such claims.

Another notable effort to articulate a unified perspective on PP is due to Stephen Gardiner (2006, 2010, 2011), who proposes a restricted version of the maximin rule as a "core" of PP. This rule recommends precaution when four conditions are satisfied: (1) nothing is known about the probabilities of the possible outcomes, (2) the precaution assures that catastrophe will be avoided, (3) the costs of enacting the precaution are minimal, and (4) any alternative action to the precaution may result in catastrophe. The thought is that, by working outwards from this core, it may be possible to attain a general understanding of PP. While I find much to admire in Gardiner's discussion, I think that his strategy for attempting to achieve a unified conception of PP is an unpromising one. The difficulty has to do with the nature of the conditions to which maximin is restricted. These conditions are such as to make the decision relatively easy. Hence, a number of decision rules that often conflict with maximin in other circumstances agree with it in the special case Gardiner examines. As a result, the "core" case provides very little indication of which direction to go when those restrictive conditions are relaxed. Should one continue to follow the maximin rule, or some other principle, and on what basis should such decisions be made? Gardiner's approach is also problematic if construed as an answer to the dilemma objection. In particular, critics assert that it is obvious that a precaution should be enacted if it is assured of preventing a potential catastrophe at practically no cost but claim that this observation is unhelpful since real issues of environmental policy generally involve hard trade-offs (see Sunstein 2005, p. 112). Thus, although Gardiner's restricted maximin interpretation of PP avoids incoherence, it is arguably skewered on the horn of triviality.[6]

Finally, let us turn to the third challenge concerning the relationship between PP and policy-relevant science. Although PP is most commonly discussed as a decision rule, it is not unusual for advocates to propose that it also has methodological implications for policy-relevant scientific research, sometimes under the banner of "precautionary science" (Barrett and Raffensperger 1999; Kriebel *et al.* 2001; Sachs 2011; Tickner and Kriebel 2006). One such implication of PP is the rejection of the ideal of value-free science. According to the value-free ideal, scientific research should be kept as separate as possible from ethical and political value judgments that inevitably influence policy decisions on environmental and human health issues (Douglas 2009; Lacey 1999; Proctor 1991). An epistemic PP conflicts with the value-free ideal by suggesting that the aims of protecting human

[6] See section 3.3.2 for a more detailed discussion of Gardiner's proposal.

health and the environment can legitimately influence methodological decisions in policy-relevant science. For example, it suggests that what should count as sufficient evidence that a new technology does not pose undue risks reflects a value judgment concerning the relative costs of unnecessary regulation versus harmful environmental or human health impacts. The argument from inductive risk is one classic and influential critique of the value-free ideal that is motivated by value judgments such as these (see Braithwaite 1953; Churchman 1948; Cranor 1993; Douglas 2009; Hempel 1965; Lemons, Shrader-Frechette, and Cranor 1997; Nagel 1961; Rudner 1953; Shrader-Frechette 1991; Steel 2010). According to this argument, the decision to accept a hypothesis involves a value judgment about what should count as sufficient evidence, a judgment that may depend on ethical considerations about the seriousness of distinct types of error. Versions of the argument from inductive risk are often encountered in discussions of the epistemic implications of PP (John 2007, 223; Kriebel *et al.* 2001, pp. 873–4; Peterson 2007, pp. 7–8; Sachs 2011, pp. 1302–3; Sandin *et al.* 2002, pp. 294–5).[7] However, these proponents of epistemic precaution do not engage with the philosophical literature criticizing the argument (Dorato 2004; Jeffrey 1956; Lacey 1999, 2004; Levi 1960, 1962, 1967; McMullin 1982; Mitchell 2004). The most common objection is that the argument from inductive risk relies on an outmoded behaviorist conception of acceptance, according to which to accept a hypothesis is to undertake some act that would be appropriate if the hypothesis were true. A defense of an epistemic PP, then, requires answering such charges. In addition, rejecting the value-free ideal requires proposing some alternative standard for distinguishing between legitimate and illegitimate influences of values in scientific research. An emerging literature on this topic exists in philosophy of science (see Douglas 2000, 2009; Elliott 2011a, 2011b, 2013; Elliott and McKaughan 2014; Kitcher 2001, 2011; Kourany 2010; Longino 2002; Steel 2010; Steel and Whyte 2012) but again has mostly been neglected in discussions of epistemic aspects of PP.

So I claim that the three challenges described above – the dilemma objection, the multitude of potentially conflicting ideas associated with PP, and the relation between PP and policy-relevant science – remain live concerns. This book is written with the firm conviction that they are not independent. Answering these challenges requires carefully examining how the several elements of PP interconnect with one another.

[7] In fact, Sandin *et al.* (2002, pp. 294–5) quote Rudner's classic (1953) statement of the argument.

In the next section, I sketch the outlines of how I believe this can be done.

## 1.2 Putting the pieces together

Questions about the interpretation of PP are an example of a classic type of philosophical puzzle. One is faced with an important, interesting, and yet tantalizingly unclear concept, and the problem is to provide a clear and coherent account of that concept, its rationale, and its logical implications. In the abstract, this is just the sort of puzzle that arises repeatedly in the Platonic dialogues with such questions as "What is justice?" or "What is love?" And as with any kind of puzzle, a solution depends on two factors: having the right pieces and putting them together in the right way. So what are the pieces to this puzzle, and how should they be assembled?

Paradigm applications of PP involve a trade-off between short-term gain, often for an influential party, against a harm that is uncertain or spatially or temporally distant. I propose that PP recommends the following three "core themes" for such decisions:

1. **The Meta-Precautionary Principle (MPP)**: The MPP asserts that uncertainty should not be a reason for inaction in the face of serious environmental threats. This principle is called "meta" because it is not a rule that indicates which of several environmental policy options to select – for instance, whether to set the allowable level of arsenic in drinking water at 50, 10, or 5 parts per billion. Instead, it places a restriction on what sorts of rules should be used for that purpose, namely decision rules that are susceptible to paralysis by scientific uncertainty should be avoided. In my interpretation, MPP is the most fundamental piece of PP insofar as it imposes constraints on the operations of the other two elements.
2. **The "Tripod"**: The term "tripod" refers to the knowledge condition, harm condition, and recommended precaution involved in any application of PP (see Trouwborst 2006). Like several other authors (Manson 2002; Munthe 2011; Sandin 1999), I take the elements of the tripod to be adjustable rather than fixed.[8] This means that there are

[8] Sandin (1999) adds a fourth element, namely how strongly the precaution is recommended – for instance, whether it is mandatory or merely permissible. In my approach, degrees of obligation have to do with the extent to which, in a particular context, adherence to PP is required. In other words, I think it is helpful to distinguish between (a) what PP recommends in a context and (b) whether one should act in accordance with PP in that context. Clearly, answering (a) is prerequisite for answering (b).

multiple ways of specifying the knowledge condition, harm condition, and recommended precaution, and careful consideration of the particulars of each application are relevant to deciding how to fill in the blanks. I will use the expression *version of PP* to refer to any statement according to which satisfying a specific knowledge and harm condition is sufficient to justify a specific precaution. For example, "If it is possible that an activity might lead to irreversible harm, then that activity should be banned" is one version of PP, while "If there is some scientific evidence that an activity will lead to irreversible harm, then an alternative should be substituted for that activity if feasible" is another. Which version of PP should be used in a given application is influenced by MPP – we should avoid versions that turn scientific uncertainty into paralysis – as well as by the next and final component.

3. **Proportionality**: Roughly, proportionality is the idea that the aggressiveness of the precaution should correspond to the plausibility and severity of the threat. I propose that proportionality be defined more precisely in terms of two subsidiary principles that I call consistency and efficiency. Consistency requires that the precaution not be recommended against by the same version of PP that was used to justify it. Efficiency states that, among those precautions that can be consistently recommended, the less costly should be preferred. Consistency and efficiency place important constraints on what can be justified by PP in a given context. For example, if there is no version of PP that can consistently recommend an action (say, preemptive war) in a given context (say, a US invasion of Iraq in 2003), then PP cannot justify that action in those circumstances. Finally, MPP affects how proportionality is applied. For example, comparisons of the relative efficiency of policy options should not be done in a way that makes scientific uncertainty grounds for continual delay.

Although these three core themes include many familiar elements of PP, as they should, the resulting proposal is distinctive in several important respects.

The most fundamental distinctive feature of the interpretation proposed here is the extent to which it ties together aspects of PP that are usually treated as separate or even conflicting. To explain this point more fully, it will be helpful to explicitly consider three possible ways of characterizing the role of PP (see Ahteensuu and Sandin 2012, pp. 971–2). One could view PP as either a

- *procedural requirement* that places some general constraints how decisions should be made; or a

- *decision rule* that aims to guide choices among specific environmental policies; or an
- *epistemic rule* that makes claims about how scientific inferences should proceed in light of risks of error.

In this categorization, MPP is an example of a procedural requirement.[9] It places general constraints on the sorts of decision rules that should be used, but it does not specify which policy option should be chosen. Decision rules, by contrast, are designed to select among specific policy options, for instance, whether to pursue efforts to reduce greenhouse gas emissions for the purpose of mitigating climate change and, if so, by what mechanisms and how aggressively. While MPP alone is not capable of indicating which policies to choose, the three components of PP in my interpretation – MPP, the tripod, and proportionality – jointly function as a decision rule. This point is illustrated in Chapter 2 in relation to climate change mitigation. Finally, epistemic rules say something about what standards of evidence should be employed in specific areas of scientific research. An example of PP construed as an epistemic rule would be the claim that toxicity assessments of chemicals should err on the side of protecting human health and the environment.

According to the interpretation proposed here, PP is all of the above: a procedural requirement, a decision rule, *and* an epistemic rule. The important thing is to grasp how these various aspects of PP work together to form a dynamic and logically coherent system. A procedural requirement such as MPP cannot be considered in isolation from the decision rules upon which it imposes constraints. Conversely, developing an adequate decision rule interpretation of PP requires being sensitive to the fundamental demand that uncertainty not be transformed into justifications of interminable delay in the face of serious dangers. Moreover, epistemic considerations are highly significant for how PP functions as a whole because of the prominent role of the concept of scientific uncertainty. In Chapter 5, I argue against the standard decision-theoretic contrast between uncertainty and risk, according to which risk entails knowledge of possible outcomes of actions and their probabilities while uncertainty means knowledge of possible outcomes but not their probabilities. Instead, I define scientific

[9] Ahteensuu and Sandin (2012, p. 972) also list legal requirements of pre-market safety testing (e.g., of chemicals) as examples of procedural requirements. However, such proposals can also be viewed from an epistemic perspective, as they express the judgment that certain materials should not be presumed sufficiently safe for commercial use until evidence is provided to that effect. Such judgments clearly reflect a relative weighting of errors as emphasized in the argument from inductive risk. Hence, I regard pre-market safety testing requirements as applications of an epistemic PP.

uncertainty as the absence of a model whose predictive validity for the task in question is empirically well confirmed. Given this proposal, knowledge of probabilities does not necessarily eliminate scientific uncertainty, which entails among other things that MPP may be relevant to cases in which quantitative estimates of probabilities exist. An examination of epistemic issues, therefore, has important implications for the scope of PP. That result is reinforced by the critique of the value-free ideal undertaken in Chapters 7 and 8. Rejecting the value-free ideal entails that PP should not be restricted to cases of unquantifiable hazards. That is because what is quantifiable and what is not depends on decisions about what methods of quantification are accepted, and an epistemic PP can influence those decisions if the value-free ideal is rejected.

Unlike the approach I advocate, interpretations of PP typically emphasize one of these three aspects – procedural requirement, decision rule, or epistemic rule – to the neglect or outright exclusion of the others. Such a fragmented approach has, in my view, a number of undesirable consequences. My own experience of the literature on PP is of a whole that is disappointingly less than the sum of the parts. Although one finds many interesting insights and proposals related to PP, it is often unclear how these pieces fit together in a logically consistent manner. Do the several distinct approaches to PP constitute separate, possibly incompatible, principles? And if they are compatible, how do they work together in a coherent fashion to inform environmental policy decisions? Moreover, fragmentation makes it difficult to adequately answer objections and to provide a well-grounded positive motivation for the principle. Let us consider some examples to illustrate these general observations.

One common approach to PP construes it as a decision rule, dismisses procedural requirement interpretations as too weak, and does not pay much attention to epistemic interpretations. For example, Gardiner criticizes the claim that scientific uncertainty should not be a reason for inaction in the face of serious threats (i.e., MPP) as being close to vacuous (2006, p. 44). Likewise, Carl Cranor calls such interpretations of PP "wimpy," because they are too easy to satisfy and are something no one would disagree with (Cranor 2004, p. 261). Katherine McKinnon refers to them as "impotent" (2009, p. 190). And Sandin (2006, p. 177) describes them as not very demanding.[10] Each of these authors then quickly drops MPP and moves on to advance one or another decision rule interpretation of PP (Cranor

[10] So-called "weak" interpretations of PP are formulated in several different ways (see Ahteensuu and Sandin 2012, pp. 970–1). One type of formulation, in the style of the 1992 United Nations Conference on Sustainability, claims that scientific uncertainty should not be a reason for delay

2001, p. 322; Gardiner 2006, pp. 45–7; McKinnon 2009, pp. 190–1; Sandin 2006, pp. 179–80).

As part of the reply to the dilemma objection discussed in Chapter 2, I show that claims about the vacuity of MPP are not true. Moreover, dismissing MPP is a seriously problematic strategy for one who wishes to defend PP. Most obviously, it amounts to immediately conceding the first horn of the dilemma objection, which charged that weak formulations of PP are vacuous. Furthermore, since MPP is crucial to the central motivation of PP (Hansson 1999), this unnecessary retreat makes it difficult to explain the principle's positive rationale. In the positive argument given for PP in Chapter 4, a central concern is to explain why environmental policy should emphasize avoiding paralysis by analysis, as MPP does. Why not emphasize the perils of rushing into environmental regulation on the basis of inadequate evidence instead? In Chapter 4, I answer this question by arguing that the history of environmental policy shows that the former type of error – delay in the face of serious danger – has been far more prevalent and serious than the opposite – environmental regulations targeting threats that were subsequently found to be nonexistent. But such an argument presumes that MPP makes a substantive and important claim. In addition, if MPP is dismissed as empty, it is very difficult to explain how PP differs from decision approaches with which it commonly contrasted, such as cost–benefit analysis. If MPP is already satisfied by cost–benefit analysis and every other seriously considered decision-making framework, then what could be the point of proposing an alternative, supposedly more precautionary, decision-making approach?

Some interpretations acknowledge MPP but treat it as relatively incidental in comparison to PP as a decision rule. In such interpretations, MPP mainly enters as a way to make the point that, in applications of PP, one should not set the bar for the knowledge condition too high as this could lead to an inability to justify any precaution whatsoever (Hartzell 2009; Munthe 2011). Although such observations are correct as far as they go, limiting MPP in this way tends to result in missing two crucial aspects of PP.

in the face of serious or irreversible threats to the environment. This is the form considered in Gardiner (2006) and Cranor (2004). Another formulates the weak PP as saying that when confronted with scientifically uncertain yet serious or irreversible threats to the environment, it is permissible to take action (McKinnon 2009; Soule 2004). While differing verbally, it is easily seen that these two formulations are logically equivalent. For if it is permissible to take action when confronted with scientifically uncertain yet serious or irreversible environmental hazards, then scientific uncertainty alone is not a sufficient reason for inaction in such circumstances. Conversely, if scientific uncertainty is not sufficient to justify inaction in the face of environmental threats, then such action must sometimes be permissible despite uncertainty.

First, when MPP is relegated to such a minor role, its central importance to the rationale of PP is easily overlooked. To give a positive rationale for PP, it is necessary to explain why environmental policy should emphasize avoiding lengthy delays due to scientific uncertainty (as MPP does). Second, to restrict MPP to a mere admonition not to insist upon an excessively strict standard of scientific proof is to fail to appreciate its implications for other aspects of PP. In particular, MPP is crucially important for how proportionality and its twin components, consistency and efficiency, are understood. For example, Hartzell-Nichols (2012) insists upon a very stringent special case of consistency in which no precaution is allowed unless it is fully certain that it will not lead to catastrophe. When the harm condition is catastrophe, satisfying Hartzell-Nichols' requirement is sufficient, *but not necessary*, for satisfying consistency. That it is not necessary is absolutely crucial from the perspective of MPP, since the demand for full certainty that the precaution could lead to no catastrophic side effects is an obvious invitation to transform uncertainty into endless delay.

More recently, an opposite approach has emerged, namely to reject PP as a decision rule and to promote it either as a procedural constraint or an epistemic rule (Peterson 2006, 2007; Sprenger 2012; Steele 2006). This approach appears to be inspired by a pair of articles claiming to provide a logical proof that PP must be incoherent if construed as a decision rule (see Peterson 2006, 2007). In section 2.5.2, I explain why the theorem in question does not in fact have the import that has been attributed to it. Moreover, just like attempting to defend PP while rejecting MPP, jettisoning PP as a decision rule is self-defeating. If PP construed as a procedural constraint has some significance, then that is because it often conflicts with aspects of decision rules conventionally used to guide environmental policy, such as cost–benefit analysis. Yet if that is the case, MPP cannot simply be grafted on to such approaches. Moreover, if new decision approaches are recommended in the name of PP – such as robust adaptive planning – then it is unclear why PP as a decision rule is said to be incoherent. Similarly, an "epistemic but not decision rule" interpretation of PP is unpromising considered on its own terms, since it is unclear how PP could be incoherent for decisions in general without also being incoherent with respect to decisions pertaining to epistemic matters.

So, the overarching theme of the interpretation advanced here is that integrating meta, decision, and epistemic aspects of PP is crucially important for a more adequate understanding of the principle. But other more specific contributions occur throughout the book. Prominent among these is the concept of consistency, which requires that the version of PP used to justify a precaution not also recommend against that precaution. Although several interpretations of PP incorporate ideas related to consistency (see Gardiner 2006; Hartzell-Nichols 2012; Munthe 2011), in each case these are less general than consistency in ways that significantly undermine their effectiveness for addressing basic challenges raised in the previous section.[11] Consistency plays an important role throughout this book. In Chapter 2, it is a crucial component of a more adequate response to the second horn of the dilemma objection that strong forms of PP are incoherent, and hence, irrational. In Chapter 3, it plays an important role in explaining how a more unified understanding of PP is possible, and in Chapter 4 it is relevant to the discussion of risk trade-off analysis. The Appendix develops a framework for formalizing the concept of consistency and uses this framework to define several additional concepts and to derive a few fundamental results. The most important result is a uniqueness theorem showing that no two fully adequate versions of PP can generate conflicting recommendations. Other specific contributions include a development of the historical argument for PP against several objections challenging its logical cogency in Chapter 4 and a new definition of scientific uncertainty proposed in Chapter 5. In addition, Chapter 6 develops a new argument for intergenerational impartiality that aims to avoid the challenge posed by agent-relative ethics. Chapter 7 defends the argument from inductive risk against the charge of behaviorism by explaining how it can be defended from the perspective of a theory of acceptance due to Jonathan Cohen (1992). Chapter 7 also defends the distinction between epistemic and non-epistemic values. That distinction is used as a basis for the values-in-science standard, which I propose as a replacement of the rejected value-free ideal. According to the values-in-science standard, non-epistemic values should not conflict with epistemic values in the design or interpretation of scientific research that is practically feasible and ethically permissible. Chapter 8 illustrates and defends the values-in-science standard in connection with a case study of uncertainty factors. This chapter also leads to the result, noted above, that PP cannot be restricted to examples involving unquantifiable probabilities because an epistemic PP

[11] See section 2.5.1 for a discussion of consistency in relation to Munthe (2011), and section 3.3 for discussions of consistency in relation to (Gardiner 2006) and Hartzel-Nichols (2012). Sandin *et al.* (2002, p. 294) come close to stating consistency in its full generality when they write, "the precautionary principle should be applied also to the precautionary measures prescribed by the precautionary principle itself." However, this statement occurs as a stand-alone sentence without further elaboration or application. The EU's Communication on the Precautionary Principle (EU 2000) defines the term "consistency" in a somewhat different way as a requirement of uniform application across cases.

is relevant to decisions about which methods of quantification are acceptable and which are not. Finally, the concluding Chapter 9 ties the threads together and applies the interpretation advanced here to three relatively brief case studies, each of which is chosen to foreground a specific aspect of PP.

CHAPTER 2

# *Answering the dilemma objection*

## 2.1 Introduction

Perhaps the most commonly voiced objection to PP takes the form of a dilemma: the principle can be given either a weak or a strong interpretation, and in the first case it is trivial and in the latter it is irrational (Burnett 2009; Clarke 2005; Engelhardt and Jotterand 2004; Goklany 2001; Graham 2001; Harris and Holm 2002; Manson 2002; Marchant and Mossman 2004; Powell 2010; Soule 2004; Sunstein 2001, 2005; Taverne 2005; Turner and Hartzell 2004). On the one hand, PP would be trivial if it merely claimed that full certainty is not a precondition for taking precautions, since this is something that every account of rational decision making already accepts. On the other, PP would be irrational if it asserted that no activity with a non-negligible risk of harm is permissible. For in that case, PP would often prohibit the same precautionary measures it prescribes, as those measures themselves sometimes come with risks of harmful consequences. In this chapter, I defend PP against this objection and argue that both horns of the dilemma are unsound.

A proper consideration of the issue requires a clarification of the relationship between so-called "weak" and "strong" versions of PP. I propose that "weak" versions of PP are not rules that are capable of selecting among alternative environmental policies. Instead, they are meta-rules that place constraints on what types of decision rules should be used, advising policy makers to avoid decision procedures that enable scientific uncertainty to become a reason against action in the face of serious potential environmental harms. So-called "strong" versions of PP, by contrast, are decision rules designed to satisfy the requirements of the meta-precautionary principle. Thus, I use the term meta-precautionary principle (MPP) in place of "weak" PP and PP instead of "strong" PP. Given this preliminary clarification, let us return to the two horns of the dilemma.

I argue that MPP is not trivial because it often recommends against some decision rules that are seriously defended in the sphere of environmental policy, including cost–benefit analysis approaches advocated by many critics of PP. Consequently, the first horn of the dilemma is mistaken: MPP is a substantive and informative proposition. Although this argument for the substantive nature of MPP may seem relatively straightforward, it is frequently overlooked by advocates of PP who quickly grant to critics the triviality of claims about scientific uncertainty not being a sufficient reason for delay in the face of environmental threats (Cranor 2004; Gardiner 2006; McKinnon 2009; Sandin 2006; Sandin *et al.* 2002). Such concessions are mistaken as well as bad tactics for one who wishes to defend PP. They are mistaken because MPP is not trivial, for the reason just noted above and elaborated in further detail in section 2.3. They are bad tactics because they undermine the position they seek to defend: if MPP is trivial, then the basic rationale and purpose of PP is called into question. Advocates of PP have often fallen into this error, I suspect, because they tend to consider MPP as a freestanding interpretation of what PP asserts rather than as one component of a more complex framework. Yet a meta-principle can only be understood in relation to the ground-level principles upon which it imposes constraints. So-called "strong" interpretations of PP, then, are decision rules designed to abide by the strictures of MPP.

That brings us to the second horn of the dilemma, which claims to demonstrate the incoherence of strong interpretations of PP. I answer the second horn of the dilemma by developing a new explication of the concept of proportionality. To state the concept of proportionality that I propose, it is necessary to introduce the idea of *a version of PP*. A version of PP claims that satisfying a specific knowledge and harm condition is sufficient justification for enacting a particular precaution. The three elements of such a claim – knowledge condition, harm condition, and recommended precaution – can be adjusted to create a large number of alternative versions of PP. Proportionality places constraints on which versions of PP may be used to justify which precautions in particular contexts. More specifically, proportionality is comprised of two principles I call *consistency* and *efficiency*. Consistency requires that the version of PP used to justify the precaution not also recommend against that same precaution. Efficiency requires that, among policy options that can be consistently recommended by the version of PP being used, those with less in the way of harmful effects should be preferred. Then I use a detailed case study concerning climate change to show how PP, applied in a proportional manner, is coherent, thereby undermining the second horn of the dilemma.

Finally, in section 2.5, I address two possible misunderstandings of the interpretation of PP advanced here and one potential objection. The two misunderstandings mistakenly identify the concept of proportionality as defined here – and consistency in particular – with other more familiar but less adequate ways of answering the charge that PP leads to incoherence. The objection concerns a theorem allegedly establishing the impossibility of any coherent interpretation of PP as a decision rule.

## 2.2 The dilemma

The charge that PP, depending on how it is interpreted, is either vacuous or obviously mistaken has been made in a variety of ways by numerous authors. Perhaps the most elaborate exposition of this dilemma is given by Sunstein (2001, 2005), who distinguishes weak (and trivial) from strong (and incoherent) formulations of PP. Weak formulations of PP assert that scientific certainty of impending harm should not be a precondition for precaution. Regarding such propositions, Sunstein writes: "The weak versions of the Precautionary Principle state a truism – uncontroversial in principle and necessary in practice only to combat public confusion or the self-interested claims of private groups demanding unambiguous evidence of harm, which no rational society requires" (2005, p. 24). Having thus disposed of weak forms of PP as trivial, Sunstein proceeds to the second horn of the dilemma. This begins with a statement of a strong form of the principle.

> For the moment let us understand the principle in a strong way, to suggest that regulation is required whenever there is a possible risk to health, safety, or the environment, even if the supporting evidence remains speculative and even if the economic costs of regulation are high. To avoid absurdity, the idea of "possible risk" will be understood to require a certain threshold of scientific plausibility. (2005, p. 24)

The fundamental problem with strong formulations of PP, Sunstein argues, is not that they disregard economic costs or that they are too vague. Instead, "The real problem is that the principle offers no guidance – not that it is wrong, but that it forbids all courses of action, including regulation. It bans the very steps that it requires" (2005, p. 26). The reason for this is that regulations that aim to protect against some potential environmental hazard may themselves pose dangers to human health or the environment. For example, a regulation prohibiting genetically modified crops might result in reduced food production; measures to slow global warming may result in the impoverishment, and consequently impaired health, of some people; a ban on nuclear power could result in heavier reliance upon coal-burning power plants and hence in more air pollution (Sunstein 2005, pp. 27–32). In general, the incoherence charged in the second horn of the dilemma arises when PP can, in a given set of circumstances, recommend both for and against the same action. Therefore, Sunstein concludes that PP is "literally incoherent" and that it is "paralyzing" instead of protective (2005, pp. 4, 34).

Several variations of the second horn of the dilemma objection exist. For example, the "absolutist" objection (Sandin *et al.* 2002, pp. 290–1) is an extreme version of the second horn of the dilemma. The absolutist objection interprets PP as banning any action that might possibly be harmful and observes that such a position is incoherent because it prohibits everything. However, the second horn of the dilemma as presented by Sunstein is more sophisticated than this, as it includes an evidential threshold as part of strong forms of PP and does not assert that an absolute ban is the only possible precaution. Another variant of the second horn of the dilemma, called the "risk trade-off" argument (see Sandin 2006, p. 177), claims that a precaution may generate risks equally or more serious than the target risk it seeks to prevent. Thus, the risk trade-off argument is naturally understood as claiming that strong versions of PP are incoherent, as Sunstein in fact does (see 2005, pp. 32–3; cf. Graham and Wiener 1995).

However, some regard risk trade-off and incoherence as separate objections. In particular, Sandin states that the risk trade-off argument does not aim to show that PP is incoherent, only that it is irrational (2006, p. 177; cf. Sandin *et al.* 2002, pp. 292–4). The relationship between PP and risk trade-off will be discussed in greater detail in Chapter 4. However, at present I merely note that it is unclear how the risk trade-off argument could show that PP is irrational without also showing it to be incoherent. For the risks generated by a precaution are either less serious than the target risk or they are not. If they are less, then the risk trade-off argument provides no reason to think PP is irrational, as it can be reasonable to run a less significant risk to avoid a more serious one. On the other hand, if risks generated by the precaution equal or exceed the target risk, then the conditions needed to show incoherence would appear to be fulfilled.

The dilemma objection is one of the most frequently encountered criticisms of PP. Many critics draw a distinction between weak but toothless and strong but unreasonable formulations of PP (Burnett 2009; Clarke 2005; Manson 2002; Marchant and Mossman 2004; Powell 2010; Soule 2004; Turner and Hartzell 2004). Sunstein's claim that strong forms of PP are incoherent is very similar to what some critics call "the precautionary paradox," according to which PP generates contradictory results through prohibiting technologies that have the potential to improve human health or well-being (Clarke 2005; Engelhardt and Jotterand 2004; Goklany 2001; Graham 2001; Harris and Holm 2002; Manson 2002; Turner and Hartzell 2004). Moreover, some critics claim that the dilemma demonstrates that PP is politically appealing only because of its ambiguity, which allows proponents to shift back and forth from weaker to stronger versions of it (Marchant and Mossman 2004, p. 14). In sum, it is a – if not *the* – central objection to PP.

## 2.3 A meta-decision rule

In this section, I address the first horn of the dilemma targeting the so-called "weak" PP, which I suggest is better characterized as the meta-precautionary principle (MPP). A meta-rule places constraints on what sorts of decision rules are acceptable but does not specify which reasons should determine environmental policy decisions or how and, therefore, does not constitute a rule in its own right for selecting among alternative environmental policies. The constraint imposed by MPP is that scientific uncertainty should not be a reason for inaction (i.e., failure to take precaution) in the face of serious environmental threats. This constraint has implications for what decision rules should be used in environmental policy making. For instance, MPP recommends against both of the following rules:

- Precaution is warranted only if it can be shown that the expected benefits of the precaution outweigh its expected costs.
- Precaution is warranted only if it is certain to prevent the threatened harm.

The first rule makes it impossible to justify precautions when expected costs and benefits cannot be reliably forecast due to scientific uncertainty. This is the argument against cost–benefit analysis developed in this section. The second rule leads to inaction when there are some possible scenarios in which no precaution can prevent the threatened harm (i.e., when it is uncertain whether the harm is preventable). This is the argument elaborated on in section 3.3.3 against the maximin rule, which asserts that one should choose the option with the least bad worst-case outcome. What MPP asserts depends in part on how "scientific uncertainty" is understood. The concept of scientific uncertainty is explored in Chapter 5, where I propose that uncertainty about a decision means a lack of knowledge that would enable its consequences to be predicted. More precisely, I propose that outcomes of actions are scientifically uncertain when no model exists whose predictive validity regarding their outcomes has been empirically well confirmed.

To show that MPP is *not* trivial, it suffices to show that it conflicts with some decision-making procedure that is actually advocated or used for evaluating environmental regulations. Consider this point in relation to cost–benefit analysis. As its name suggests, cost–benefit analysis attempts to predict the costs and benefits of proposed policy options, in monetary terms, and to identify the option that has the greatest margin of benefits over costs (Jaeger 2005, pp. 8–10). Cost–benefit analysis is often presented as a procedure for deciding among policy options (Frank 2005; Lomborg 2010), and the idea that cost–benefit analysis should be used as a test for proposed regulations is also suggested in Executive Orders issued by the Reagan and Clinton Administrations (orders 12991 and 12866, respectively). And many critics of PP defend cost–benefit analysis as a basis for environmental policy (see Graham 2001; Marchant 2001a; Posner 2004; Sunstein 2001).

Yet cost–benefit analysis can easily conflict with the MPP's demand that scientific uncertainty not be a reason for failure to take precautions against serious threats to human health or the environment. This point is nicely illustrated by attempts to estimate the social cost of carbon in climate change economics. The social cost of carbon (SCC) "is the total damage from now into the indefinite future of emitting an extra unit of GHGs now" (Stern 2007, 28). SCC is of crucial importance from a cost–benefit perspective because it is the key indicator of how much or how little should be spent on climate change mitigation, that is, on efforts to reduce greenhouse gas (GHG) emissions. The difficulty of judging whether the benefits of substantial climate change mitigation would exceed or fall short of their costs, then, can be illustrated by noting the extraordinarily wide range of estimates of SCC. A review published in 2005 found 103 estimates of SCC varying from $0 to $273 per ton of $CO_2$ (Tol 2005).[1] To get a better sense of the differences, consider three widely discussed climate change mitigation cost–benefit analyses.

- Nicholas Stern (2007, p. 322): SCC for BAU (business as usual) estimated at about $85 per ton of $CO_2$. Recommends aggressive mitigation efforts to stabilize atmospheric $CO_2$ equivalent concentrations at between 450 to 550 ppm; recommends a cap-and-trade scheme,

[1] The emissions scenarios used to generate these estimates varied although most were intended to represent a "business as usual" (BAU) situation.

government support for research on low-carbon technologies, and action to reduce deforestation (2007, pp. xvi–xviii).

- **William Nordhaus** (2008, p. 91): SCC for baseline scenario at about \$7.50 per ton of $CO_2$. Recommends a global harmonized carbon tax set at approximately \$9.30 per ton of $CO_2$ in 2010, \$11.50 in 2015, \$24.50 in 2050, and \$55 in 2100, which is expected to limit $CO_2$ concentrations to 586 ppm by 2100 and 659 ppm by 2200 (Nordhaus 2008, p. 103).[2]
- **Richard Tol** (2010, 90–5): SCC for BAU estimated at approximately 55 cents per ton of $CO_2$. Proposes that the only reasonable mitigation measure would be a global harmonized carbon tax set at that rate. Tol's analysis appears in a volume edited by Bjørn Lomborg, which ultimately reaches the conclusion that mitigation should not be a focus of climate change policy (Lomborg 2010, pp. 395–6).

To get a sense of the practical implications of these differences, consider what each of the SCC estimates given above would mean if converted directly into a tax on the carbon content of gasoline, adjusting for inflation to 2010 dollars in each case. Stern's \$85 SCC would correspond to approximately a \$1 per gallon gasoline tax. Nordhaus's "ramp" would translate to gasoline taxes of approximately 10 cents in 2010, 13 cents in 2015, 27 cents in 2050, and 61 cents in 2100. Finally, Tol's proposal would mean a gasoline tax of about 0.6 cents per gallon.

In sum, cost–benefit analysis requires an ability to make quantitative predictions of the costs and benefits of alternative policy options, but scientific uncertainty can undermine the ability to make such predictions in a non-arbitrary way. Scientific uncertainty, then, can easily lead to an inability to decide whether the benefits of a regulation would be greater or less than the costs. In such circumstances, cost–benefit analysis as the central basis for decision making can result in paralysis, since no action can be unambiguously justified in its terms (see Ackerman 2008a; Ackerman and Heinzerling 2004; Gardiner 2011, chapter 8; Mitchell 2009, pp. 87–9). Consequently, MPP recommends against cost–benefit analysis as a general basis for environmental decision making.

One possible response for an advocate of cost-benefit analysis is to suggest that in a state of pure uncertainty – wherein *nothing at all* is known about which scenario is more plausible than another – one should apply the principle of indifference and assign equal probabilities to all

[2] Both Nordhaus and Tol state their estimates in units of dollars per ton of carbon rather than per ton of $CO_2$. The per ton carbon estimates are easily converted to per ton $CO_2$ given the atomic weights of carbon and oxygen.

outcomes (see Bognar 2011). Even putting aside well-known logical difficulties confronting the principle of indifference (Salmon 1966), this reply does not undermine the argument that cost–benefit analysis is susceptible to paralysis by scientific uncertainty. That is because the uncertainty in environmental issues is very rarely pure. Typically, qualitative knowledge exists that would enable some discriminations to be made between more and less serious possibilities, but that knowledge does not suffice for accurate quantitative long-term predictions, as the example of climate change illustrates.

A different approach would be to combine cost–benefit analysis with a decision-theoretic approach that considers sets of probability distributions rather than a single one. Several decision rules have been discussed for such contexts.[3] One approach is to weight the possible probability distributions according to their reliability and then select the action that has the greatest reliability-weighted expected utility. Estimates produced by the US Interagency Working Group on the Social Cost of Carbon illustrate this type of procedure (Interagency Working Group 2013). Their SCC estimates are the average results of three Integrated Assessment Models (IAMs) – in fact, the same models used by Stern, Nordhaus, and Tol.[4] Thus, the Interagency Working Group's approach implicitly places zero weight on all models not included in the analysis and equal weight on the three IAMs that are included. However, approaches incorporating some version of reliability-weighted expected utility do not effectively address the challenge to cost–benefit analysis considered in this section, because quantitative reliability weights are likely to be highly uncertain as well – and perhaps even more uncertain than the original probabilities. In these circumstances, any choice of weighting is ultimately arbitrary. That point is illustrated by the Interagency Working Group's analysis. Why should total weight be placed on the three chosen IAMs when others appear at least as plausible, including some that would generate much higher estimates of SCC (see Ackerman and Stanton 2013; Frisch 2013)? And why should all IAMs included in the model be weighted equally when there may be reasons for regarding some as more plausible than others (see Ackerman and

[3] See Hansson (1994, chapter 8) for a concise overview of this literature.

[4] Although the Interagency Working Group uses the same IAMs, their estimates differ from those of Stern, Nordhaus, and Tol due to using distinct emission scenarios and parameter settings. For instance, the Working Group uses a higher discount rate than Stern. See Chapter 6 for further discussion of future discounting and section 9.2 for further discussion of the Interagency Working Group's SCC estimates.

Munitz 2012a)? Thus, the reliability-weighted expected utility approach merely replicates the indeterminacy of ordinary cost–benefit analysis at a meta-level.

A second decision-theoretic approach to cases involving uncertain probabilities takes maximin expected utility as the appropriate criterion. In this approach, one has a set of probability distributions judged to be possible given background knowledge and then selects the act that maximizes the minimum expected utility for any possible probability distribution. However, this procedure departs from the fundamental demand of cost–benefit analysis that decisions should aim to achieve the greatest margin of benefits over costs and instead attempts to avoid the worst that is likely to happen. Furthermore, the maximin rule is problematic from the perspective of MPP as noted above and discussed in further detail in section 3.3.3.[5]

Why, then, do Sunstein and other critics think that PP can be brushed aside as a triviality? I suggest two reasons for this. The first involves an equivocation on "scientific uncertainty": MPP would be trivial *if* "scientific uncertainty" merely meant "probability less than 1." Sunstein's argument turns on interpreting "scientific uncertainty" in precisely this way. Consider this passage, which occurs in *Laws of Fear* immediately prior to the passage quoted above that dismisses weak versions of PP as uncontroversial truisms:

> Every day, people take steps to avoid hazards that are far from certain. We do not walk in moderately dangerous areas at night; we exercise; we buy smoke detectors; we buckle our seatbelts; we might even avoid fatty foods (or carbohydrates). Sensible governments regulate risks that, in individual cases or even in the aggregate, have a well under 100 percent chance of coming to fruition. An individual might ignore a mortality risk of 1/500,000 because that risk is awfully small, but if 100 million citizens face that risk, the nation had better take it seriously. (Sunstein 2005, pp. 23–4)

This passage makes the interpretation of "scientific uncertainty" as "probability less than 1" explicit. Yet I know of no serious formulation of PP that interprets "scientific uncertainty" in this manner. Nor can small probabilities be equated with unpredictability. As Sunstein's example illustrates, if the probability of an event is 1/500,000, then one could reasonably predict that it will not occur in a single case but that it will occur approximately 200 times in a sample of 100 million.

[5] Some proposals combine elements of the reliability-weighted expected utility and the maximin expected utility approaches (see Hansson 1994, chapter 8). Such combined proposals inherit the difficulties of both.

At a later point in *Laws of Fear*, Sunstein does consider the possibility that "scientific uncertainty" in MPP includes situations in which probabilities of relevant outcomes are unknown (2005, pp. 59–61). He responds to this suggestion by interpreting PP as the maximin rule and arguing that the maximin rule is problematic. But this reply is seriously flawed in two respects. First, it does not address the central point that Sunstein's argument against MPP collapses once one acknowledges that "scientific uncertainty" is not equivalent to "probability less than 1." Second, no reason is given to suppose that the maximin rule is the only or best way to interpret PP.

There is an additional reason why MPP may appear to be trivial, namely that it does not recommend any specific remedy or precaution in response to any environmental harm. Indeed, it is not unusual for defenders of PP to dismiss MPP on the grounds that it is wimpy, extremely minimal, impotent, or not very demanding (Cranor 2001, 2004; Gardiner 2011, p. 412; McKinnon 2009; Sandin *et al.* 2002). But to brush aside MPP on these grounds is to misunderstand its role as a meta-principle rather than a decision rule. In addition, abandoning MPP is a self-defeating move for one who wishes to defend PP. For to say that MPP is trivial or not too demanding or impotent or extremely minimal is to imply that all existing, seriously considered decision-making approaches already satisfy it. But if that were the case, then it would be difficult to see the point of advocating PP in the arena of environmental policy. Thus, dismissing MPP as trivial is not merely mistaken. It also hobbles the project of interpreting PP by removing its underlying motive and guiding rationale.

## 2.4 Proportionality

The second horn of the dilemma challenges PP on the grounds that it is incoherent and, hence, irrational. If applied to the case of climate change, for instance, PP recommends that efforts be taken to significantly reduce greenhouse gas emissions. But this recommendation can also be considered from the perspective of PP: perhaps substantial cuts in greenhouse gas emissions would result in a global economic depression, the rise of totalitarian dictatorships, and, finally, nuclear war (Manson 2002, p. 273). In the face of such dire possible consequences, the objection concludes, PP must surely demand that we refrain from acting, thereby contradicting its initial mandate. This section develops the concept of proportionality as an answer to the incoherence horn of the dilemma.

### 2.4.1 Consistency and efficiency

The intuition behind proportionality is expressed by proverbs such as "the cure should not be worse than the disease" or "never use a cannonball to kill a mosquito." Somewhat more formally, proportionality requires that "measures be calibrated to the degree of scientific uncertainty and the seriousness of the consequences feared" (Whiteside 2006, p. 53). Proportionality is a long-standing feature of PP. According to Trouwborst: "From the start, proportionality has been a crucial feature in the application of the precautionary principle, in the sense that precautionary responses ought to correspond to the perceived dimensions of the risks involved" (Trouwborst 2006, p. 150). For example, Germany's *Vorsorgeprinzip* of the 1970s – often viewed as the wellspring of PP – makes explicit reference to proportionality (Trouwborst 2006, pp. 151–2). The European Union's Communication from the Commission on the Precautionary Principle also lists proportionality as a general principle to be followed in applying PP (EU 2000, section 6.3.1; also see von Schomberg 2006).[6] Despite the importance of proportionality to applications of PP, however, the concept has received little in the way of explication. Consequently, advancing proportionality as answer to the second horn of the dilemma requires first proposing an account of what it is.

I approach proportionality through a consideration of the structure of PP. As several authors have noted (Cranor 2001; Manson 2002; Sandin 1999; Trouwborst 2006), PP can be thought of in terms of at least three[7] fundamental components: a harm condition, a knowledge condition (where the knowledge demanded may fall short of certainty), and a recommended precaution. Entries in each of these categories can admit of degrees: harms can be more or less severe; the knowledge supporting the existence of the harm and its relation to a particular activity may be more or less firm, and the precaution may be more or less strict. As a result, many versions of PP can be generated from the basic abstract schema of harm plus uncertain knowledge leads to precaution (Manson 2002). Consider these three examples:

1. If there is some scientific evidence that an activity leads to a significant and irreversible harm, then an alternative should be substituted for that activity if feasible.

[6] See Trouwborst (2006, pp. 149–53) for references to many other documents concerning PP that mention or discuss proportionality. In Chapter 3, I explain how the concept of proportionality is linked to alternatives assessment, which plays a prominent role in North American discussions of PP (O'Brien 2000; Raffensperger and Tickner 1999; Tickner and Geiser 2004).

[7] Sandin (1999) adds a fourth: the degree to which the precaution is mandatory.

2. If a scientifically plausible mechanism exists whereby an activity can lead to a catastrophe, then that activity should be phased out or significantly restricted.
3. If it is possible that an activity will lead to a catastrophe, then that activity should be prohibited.

Versions 2 and 3 require a more severe harm than 1 to be triggered (catastrophe versus irreversible environmental harm), whereas versions 1 and 2 both demand a more substantial body of knowledge than 3. Finally, the precautions demanded by versions 1 through 3 run from milder to stricter. Proportionality is not an extra box in this schema; rather, it has to do with how the levels of harm, knowledge, and precaution interact with the specifics of the case in question (see Trouwborst 2006, p. 151). In particular, I propose that the following two principles lie at the heart of proportionality: *consistency* and *efficiency*.

Consistency states that a precaution should not be precluded by the same version of PP used to justify it.[8] To illustrate, consider an example of an *in*consistent precaution. Imagine someone who calls for a ban of the measles/mumps/rubella (MMR) vaccine on the grounds that it may cause autism. Suppose that the person defends this position by appeal to the following version of PP:

4. If it is possible that an activity will lead to serious harm, then that activity should be prohibited.

But such an invocation of PP would fly in the face of consistency. For banning the MMR vaccine is also very likely to lead to seriously harmful effects due to increased incidence of disease. Similarly, if a particular precaution is recommended on the basis of version 2 of PP, then there should not be a scientifically plausible mechanism through which this precaution itself will lead to catastrophe. In general, consistency requires that the precaution not have effects that satisfy the harm and knowledge conditions stated in the antecedent of the version of PP being applied. I take the proverb that the cure should not be worse than the disease to be an expression of this aspect of consistency. That proverb counsels against a cure whose potential side effects are as harmful and as well grounded given our knowledge as the malady it seeks to remedy. But harmful unintended consequences are not the only concern pertinent to consistency. Ineffectual precautions can also be a problem: if the recommended precaution is ineffectual, the relevant version of PP will recommend against it for the same reason it advises against the status quo.

[8] See section 1.2 of the Appendix for a formal explication of the concept of consistency.

However, one qualification is needed here, namely that consistency restricts attention to preventable harms. In other words, a potential harm is a reason against performing an action only if there is some other action through which that harm could be avoided. This qualification is important, since otherwise no action could be consistently recommended if there were a scenario meeting the knowledge condition in which disaster occurs no matter what. That in turn would make PP susceptible to paralysis when it is uncertain whether any fully effective precaution exists, thereby leading to conflict with MPP (see section 3.3.3).

Whether a precaution can be consistently recommended by a given version of PP, then, depends on three factors: the version of PP used, the precaution recommended, and the details of the situation. A given precaution in a particular context might be consistently recommended by one version of PP but not another. For instance, a precaution might be consistently recommended by version 2 but not by version 3 of PP in which the mere possibility of catastrophe triggers an outright ban. This point is illustrated by the discussion of climate change mitigation in the next section. And a version of PP in a particular context might consistently recommend some precautions but not others. For example, version 3 could not consistently recommend that all uses of fossil fuels be completely phased out within a period of 30 days, but in the next section I make the case that it can consistently recommend the more measured sorts of mitigation policies that have actually been proposed. The MMR vaccine example illustrates the same point. While consistency would not allow the vaccine to be banned, less drastic precautions, such as replacing the preservative thimerosol with a less toxic alternative, can be consistently recommended. Similarly, while a ban on cell phones because of concerns about adverse effects of electromagnetic fields (EMFs) could not be consistently recommended, limits on the quantity of EMFs emitted by these devices could be (see section 4.2.2). Finally, case-specific details matter to what can be consistently recommended even if the precaution and version of PP are held constant. For instance, an absolute ban on a possibly toxic substance could be consistently recommended if a feasible alternative exists that is known to be entirely harmless. However, if no feasible alternative exists and the substance has some important use (such as a preservative for vaccines), then the mere possibility of adverse effects would not be sufficient to consistently justify the ban.

Efficiency demands that if more than one precaution can be consistently recommended by the version of PP being used, then those with lower costs should be preferred. Efficiency makes PP sensitive to effects that do not rise to the level of the knowledge and harm conditions present in the version of PP being applied. For example, there are a number of economic arguments for the claim that carbon taxes are a more efficient means of climate change mitigation than cap-and-trade schemes (see Hsu 2011). The adage "never use a cannonball to kill a mosquito" can be construed as an expression of efficiency and its connection to proportionality. A flyswatter will do just as well and will cause much less collateral damage. Supposing that the danger posed by the mosquito (say, the spread of a fatal disease) is worse than a hole blasted in the wall, it is efficiency rather than consistency that would recommend the flyswatter over the cannon as a means of eradicating the mosquito. Efficiency is discussed in further detail in section 3.4, while consistency is the primary focus in the following section because it is the aspect of proportionality most directly relevant to answering the incoherence horn of the dilemma objection.

An application of PP requires selecting a relevant version of PP. If there is more than one version of PP that could be consistently applied to the case, one natural approach is to identify a desired safety target that can be consistently applied and then define the harm condition as the failure to attain that target (Ackerman and Stanton 2013, p. 86). Since MPP insists that scientific uncertainty should not be a reason against precaution, the next step would be to select the least stringent knowledge condition that results in a consistently applicable version of PP given the harm condition. This approach to selecting a version of PP will be illustrated in relation to climate change mitigation in Chapter 9, and its formal details are worked out in section 1 of the Appendix. To answer the incoherence horn of the dilemma, however, it is only necessary to show that *some* version of PP exists that can be consistently applied in a realistic example to generate informative results. That is the task undertaken in the next section.

### 2.4.2 *Proportionality and climate change*

In this section, I illustrate the concept of proportionality in relation to the issue of climate change mitigation, and this case study serves as a basis for answering the second horn of the dilemma. In particular, I explain how version 2 of PP generates informative implications in this case without incoherence.

As explained above, many versions of PP can be constructed, and questions about proportionality require being clear about which version is being applied in the case at hand. For the present purposes, version 2 of PP given above will suffice: If a scientifically plausible mechanism exists whereby an activity can lead to a catastrophe, then that activity should be phased out or significantly restricted. I understand a "scientifically plausible mechanism" to be a causally related sequence of events (a) that is grounded in scientific knowledge, such as physical laws, and (b) for which scientific evidence exists of its actual occurrence. I take it as established that these conditions are met in the case of climate change (Solomon *et al.* 2007), and I think that its potential effects can be reasonably characterized as catastrophic. Here is a partial list of impacts from the 4th Assessment Report of a rise in global temperatures greater than 2 degrees Celsius.[9] Asterisks indicate levels of confidence associated with claims: *** very high confidence, ** high confidence, * medium confidence.

- **Health:** Although some risks would be reduced, aggregate health impacts would increase, particularly from malnutrition, diarrheal diseases, infectious diseases, floods and droughts, extreme heat, and other sources of risk */**.
- **Water resources:** Severity of floods, droughts, erosion, water quality deterioration will increase with increasing climate change ***. Sea level rise will extend areas of salinization of groundwater, decreasing freshwater availability in coastal areas ***. Hundreds of millions of people would face reduced water supplies **.
- **Greenland ice sheet:** Commitment to widespread ** to near-total * deglaciation, 2–7 meter sea level rise over centuries to millennia * (Parry *et al.* 2007, pp. 787–9).

These outcomes would appear to fit Hartzell-Nichols' definition of "catastrophic" as referring to circumstances "in which many millions of people could suffer severely harmful outcomes" (2012, p. 160). Furthermore, climate science since the IPCC 4th Assessment Report, issued in 2007, suggests that the IPCC (Intergovernmental Panel on Climate Change) estimates of several key outcomes, such as sea level rise, are overly optimistic. For instance, the 2007 assessment reports do not take into account a number of important feedback cycles, such as methane released by thawing permafrost (see Schuur and Abbott 2011).

So climate change appears to be a case in which the antecedent of version 2 of PP is satisfied, which then leads to the question of what precautions should be taken. The remedy specified in version 2 is that the harmful activity, in this case anthropogenic GHG emissions, "should be phased out or significantly restricted." In discussions of climate change,

[9] The *Impacts, Adaptation and Vulnerability* volume of the IPCC's 5th Assessment Report was not available at the time of this writing.

efforts to reduce GHG emissions are known as "mitigation," while the term "adaptation" refers to measures taken to adjust to adverse effects of climate change – such as rising sea levels – as they occur. Given that some harmful effects of climate change are now unavoidable, it is a foregone conclusion that resources will have to be devoted to climate change adaptation. In this chapter, I focus on the issue of mitigation. In very general terms, it is relatively clear what form mitigation would take. GHG emissions are an example of what economists call a *negative externality*: a cost generated as a by-product of market activities that is distributed across society generally rather than borne specifically by the producers and consumers of the goods involved (Nordhaus 2008; Stern 2007, chapter 2). Consequently, those implicated in generating the externality – which in the case of GHG emissions means almost all of us – have no individual incentive to change their behaviors so as to reduce its cost. In theory, the solution to problems generated by negative externalities is straightforward: introduce some mechanism whereby the previously externalized cost is directly attached to the activities that generate it. The simplest way to internalize the social costs of carbon emissions is through a carbon tax, which affixes a tax on the carbon content of fuels (Hsu 2011; Nordhaus 2008; Posner 2004). An alternative approach is a cap-and-trade scheme in which tradable emission permits are auctioned to major GHG sources, such as public utilities, factories, or large-scale agricultural operations. Cap-and-trade schemes focused on GHG emissions have been implemented in the European Union, a group of states in the northeastern United States (the Regional Greenhouse Gas Initiative), and the state of California, while examples of carbon taxes will be discussed below.

Consistency and efficiency would come into play in deciding which mitigation approach should be pursued and how its details should be worked out. In the case of a carbon tax, for example, consistency would demand that the tax not be so high or introduced so abruptly as to create an economic catastrophe, and efficiency would demand that restrictions be designed so as to achieve as much reduction as possible at the least cost. In my view, efficiency favors carbon taxes over cap-and-trade schemes (see Hsu 2011). If a carbon tax is the preferred policy option, then efficiency requires that it be implemented in a manner that avoids imposing unnecessary economic costs. For instance, proceeds from a carbon tax could be used to reduce other taxes, including tax reductions or payments specifically targeted to low-income individuals or households (Nordhaus 2008, pp. 156–8; Summer, Bird, and Smith 2009). The term "carbon tax shift" is sometimes used to emphasize that such proposals shift away from distortionary taxes (e.g., on income, employment) and toward taxes on commodities associated with significant negative externalities (e.g., carbon-based fuels). A carbon tax shift was implemented in the Canadian province of British Columbia in 2008 (Elgie and McClay 2013), and carbon tax shifts have been implemented in a number of European countries since the 1990s (Anderson *et al.* 2007). Of course, efficiency demands that the least burdensome *effective* precaution be taken, which would mean that PP would prioritize setting the carbon tax high enough to produce meaningful results, such as stabilizing atmospheric $CO_2$ equivalent at levels deemed acceptable. Given the global nature of climate change, it is plain that achieving such a goal requires that carbon reduction policies pursued by separate states be coordinated so as to attain effective results overall. This might be achieved through an international "harmonized" carbon tax (Nordhaus 2008, chapter 8).

Let us now return to the second horn of the dilemma, which claimed that "strong" versions of PP are incoherent because they ban the very actions they prescribe. Clearly, any such application of PP would not be compatible with consistency and hence would not be proportional. In other words, it would be a *mis*application of PP. Given a proper understanding of the role of proportionality in implementing PP, then, the purport of the second horn of the dilemma is rather unclear. Perhaps the aim is to show that *actual* applications of PP often fail to be consistent. Consider this in relation to the case of climate change mitigation. Showing a genuine violation of consistency requires showing that the *same* harm and knowledge conditions used to justify the *actually* recommended precaution can also be used to rule against that precaution. I will use the term *counter-scenario* to refer to a claim about how such violation of consistency might arise in a particular context. Yet the alleged counter-scenarios in the case of climate change mitigation are highly problematic. Consider Manson's self-styled "wild story" about reductions in GHG emissions resulting in worldwide economic depression, political instability, and finally nuclear holocaust (Manson 2002, p. 273). Such a scenario might deserve serious consideration if the proposed action against climate change were an immediate and absolute ban on all use of fossil fuels (see Gardiner 2011, pp. 20–1). But Manson's "wild story" utterly fails to satisfy the knowledge condition of a scientifically plausible mechanism with respect to actually proposed and implemented mitigation measures such as a carbon tax or cap-and-trade.

Similar issues arise for Sunstein's discussion of the incoherence objection in relation to climate change. According to Sunstein: "A great deal of work suggests that significant reductions in such [i.e., GHG] emissions would have large benefits; but skeptics contend that the costs of such decreases would reduce the well-being of millions of people, especially the poorest members of society" (2005, p. 27). Sunstein does not provide additional supporting details or references in the passage above, making it difficult to judge what costs he has in mind. However, in a later chapter Sunstein cites alarming estimates of economic effects of the Kyoto Accord, which he attributes to a page from the website of the American Petroleum Institute (2005, p. 173 fn. 26). Despite granting that these estimates are "almost certainly inflated" due to disregarding "technological innovations that would undoubtedly drive expenses down," he nevertheless takes them as sufficient to demonstrate the potentially adverse impacts on the poor of a carbon tax (Sunstein 2005, p. 173). But Sunstein's reasoning is difficult to understand. Why should "almost certainly inflated" estimates of the costs of the Kyoto Accord, which recommended a cap-and-trade scheme, be regarded as a basis for assessing the economic impacts of a carbon tax? Moreover, Sunstein does not consider the possibility, mentioned above, that revenues from a carbon tax could be used to reduce other taxes and to assist low-income households.[10] He also does not consider the possibility of phasing in and gradually "ramping up" a carbon tax so as to reduce economic disruptions resulting from a shift toward less GHG-intensive energy sources (see Nordhaus 2008).

There is in fact little basis for the idea that substantial climate change mitigation, for instance, by means of a carbon tax, would lead to economic catastrophe. For example, the Stern Review of the Economics of Climate Change estimates the expected costs of stabilizing $CO_2$ equivalent levels at between 500 to 550 parts per million (ppm) to between −1% to 3.5% of global GDP by 2050, with 1% as the most likely number (Stern 2007, pp. xvi–xvii, 318–21). As a comparison, about 2.2% of the world's GDP (gross domestic product) was devoted to military spending in 2011, while in the United States military spending comprised about 4.7% of GDP.[11] In the most optimistic scenario, then, climate change mitigation would stimulate economic growth in coming decades (for instance, if cost-effective alternative energy technologies quickly emerged), and in the most pessimistic scenario, the costs would still be less than what some nations choose to devote to military spending. Moreover, since climate change is very likely to eventually inflict adverse effects on the world economy, pursuing mitigation now can be expected to generate positive economic effects after about 2080 (Stern 2007, p. 321). Although some aspects of the Stern Review are

[10] In fact, Posner (2004, pp. 155–7), who is the specific target of Sunstein's criticism, proposes use of carbon tax revenue to reduce other taxes.

[11] See the Stockholm International Peace Research Institute (www.sipri.org).

controversial, the basic point about the affordability of substantial climate change mitigation is widely accepted among economists. For instance, the climate economist William Nordhaus states: "The claim that cap-and-trade legislation or carbon taxes would be ruinous or disastrous to our societies does not stand up to serious economic analysis" (Nordhaus 2012). Thomas Schelling (1997, p. 10), and even Lomborg (2001, p. 323) and Tol (2010, p. 91), who oppose substantial climate change mitigation efforts, make similar statements.

Actual experience with carbon taxes shifts support for such claims. For instance, a recent review of British Columbia's carbon tax experience concludes: "BC's [British Columbia's] carbon tax shift has been a highly effective policy to date. It has contributed to a significant reduction in fossil fuel use per capita, with no evidence of overall adverse economic impacts, and has enabled BC to have Canada's lowest income tax rates" (Elgie and McClay 2013, p. 1). As of 2012, British Columbia's carbon tax rate stands at 30 Canadian dollars per metric ton of $CO_2$ equivalent emissions. Thus, despite being approximately three times the SCC estimate of Nordhaus (2008) and nearly sixty times that of Tol (2010), British Columbia's carbon tax has had no discernible harmful economic consequences. Indeed, British Columbia's GDP slightly outpaced the rest of Canada's from 2008 to 2011, the most recent date for which GDP estimates are available (Elgie and McClay 2013, p. 5). The carbon tax shift also appears to have been effective. Use of petroleum fuels subject to the carbon tax in BC decreased by 17.4% from 2008 to 2012, while increasing by 1.5% in the rest of Canada during the same period (Elgie and McClay 2013, p. 2). Studies of carbon tax shifts in Europe have similarly found little or no adverse economic impacts. According to the *Competiveness Effects of Environmental Tax Reforms* (COMETR) report issued by the European Commission, "Overall, the net costs of ETR [Environmental Tax Reforms] are, in most sectors, exceeded by the value of the gains in energy efficiency which has been obtained over the same period of time" (Anderson *et al.* 2007, p. 67). And even for the exceptions – that is, energy-intensive industries without readily available technological means for reducing their reliance on carbon-based fuels – COMETR finds that "the burden of ETR . . . remains modest" (ibid.).

If a carbon tax shift will not result in *economic* catastrophe, might it nevertheless result in a catastrophe of some other sort? Lomborg argues that efforts to reduce GHG emissions would be a bad idea because the money would be better spent on measures to address problems that currently afflict developing nations, such as HIV/AIDS (Lomborg 2001, 2007, 2010). One possible reading of this argument is as a counter-scenario in which actions to curb climate change inadvertently lead to catastrophe in the form of millions of deaths due to disease.[12] But again, there is no scientifically plausible mechanism by which action on climate change would preclude action on these other issues or even make such action less likely (see Gardiner 2011, pp. 281–4). The lack of global action in the face of climate change in the past two decades has not been accompanied by any compensatory outpouring of poverty reduction assistance from wealthier to poorer nations (Singer 2002, pp. 23–4). So it is difficult to understand why (say) a carbon tax would lead to reductions in funding for international programs to alleviate disease and poverty – especially if the economic costs of a carbon tax shift are relatively small, as the available data suggest. Indeed, if desired, some portion of the revenue generated by a carbon tax could be put toward international poverty reduction programs. Moreover, Lomborg's argument neglects the link between climate change and global economic inequality. The early adverse effects of climate change are expected to disproportionately target the world's poor, while carbon tax revenues would be generated at a much higher per capita rate by wealthier nations. Consequently, for the near future, action to mitigate climate change could constitute a kind of "foreign aid program" (Schelling 1997, p. 8) and hence should be supported by those who are genuinely concerned about global inequalities and poverty.[13]

The counter-scenarios just considered, then, fail to demonstrate any conflict with consistency and all for the same reason. Each involves an outcome that might reasonably be described as catastrophic but in each case there is no scientifically plausible mechanism by which the precaution would lead to that outcome. Thus, in none of these counter-scenarios can the version of PP invoked in support of the precaution (i.e., version 2 from section 2.4.1) also be used to recommend against that very precaution. At best, these counter-scenarios show that a version of PP with a much *weaker* knowledge condition (e.g., version 3) would recommend against taking prompt action to curb GHG emissions. But that is no conflict with consistency, which requires that the precaution not be prohibited by the *same* version of PP that recommends it.

In sum, since applications of PP that would ban the very steps they recommend are not allowed by consistency, the second horn of the dilemma cannot be an argument against the interpretation of PP proposed here.

[12] For example, this may be what Sunstein (2005, p. 27) intends. Sunstein does cite Lomborg (2001) along with Posner (2004) and Nordhaus and Boyer (2000), but only to support the claim, "Scientists are not in accord about the dangers associated with global warming" (Sunstein 2005, p. 27).

[13] See (Ackerman 2008b) and (Zenghelis 2010) for similar critiques of Lomborg.

Moreover, the discussion in this section explains how the PP generates substantive policy guidance on the topic of climate change mitigation. For instance, the discussion explains why PP, unlike cost–benefit analysis, rejects the Tol–Lomborg position that no substantial climate change mitigation measures should be enacted and why it supports a more aggressive approach to climate change mitigation than that defended by Nordhaus (see section 9.2 for further discussion). Thus, the interpretation of PP I propose avoids both horns of the dilemma and thereby shows the dilemma to be false.

## 2.5 Two misunderstandings and an objection

In this section, I address two possible misunderstandings of my proposal and a potential objection. The first misunderstanding is to equate consistency with a *de minimis* condition requiring that evidence for the threat attain some specified threshold of plausibility before PP is triggered. The second misunderstanding is that consistency is equivalent to precluding versions of PP that systematically ban all possible options. The objection consists of a logical proof purporting to show that PP is incoherent if construed as a decision rule (Peterson 2006, 2007).

### *2.5.1* De minimis *and decisional paralysis*

A common response to the incoherence horn of the dilemma objection is to attach a *de minimis* condition to PP (see Peterson 2002; Sandin 2005; Sandin *et al.* 2002, pp. 291–2). According to this proposal, PP is appended with an across-the-board evidential threshold and is implemented only when evidence of the potential harm crosses that line. This approach effectively dismisses some counter-scenarios given to support the incoherence horn of the dilemma, such as Manson's "wild story" of climate change mitigation leading to a nuclear holocaust. However, a *de minimis* condition is not sufficient to answer the second horn of the dilemma in general, and, moreover, consistency neither entails nor is entailed by such a condition.

One objection to a *de minimis* condition is that it requires an arbitrary decision about where to set the evidential line (Munthe 2011, pp. 24–5). But my chief concern will be with another defect of the *de minimis* proposal, namely that it does not block the incoherence charged in the second horn of the dilemma. To see this, consider the following modification of version 2 of PP:

5. If a scientifically plausible mechanism exists whereby an activity can lead to a catastrophe, then that activity should be immediately and completely banned.

Version 5 differs from version 2 only in the strictness of the recommended precaution. Where version 2 calls for a phase-out or substantial restrictions, version 4 insists on an immediate prohibition. The knowledge condition is the same in both cases, and I presume demanding enough to satisfy a *de minimis* condition of the sort contemplated. However, the potential for version 5 to lead to violations of inconsistency is obvious. In the case of climate change mitigation, for instance, version 5 would recommend both for an immediate ban on fossil fuel use and against that very same ban. Thus, a *de minimis* condition does not entail consistency. In addition, consistency does not entail a *de minimis* condition. A version of PP with a very minimal knowledge condition could be consistently applied if it were known with certainty that enacting the precaution would do no harm. For example, even a very small chance of a bicycle accident is a sensible reason to wear a helmet. The interpretation of PP proposed here, therefore, does not adopt any fixed evidential standard. The choice of knowledge condition is case dependent and is driven by MPP in combination with the need for consistency.

The second possible misunderstanding of consistency interprets it as merely eliminating from consideration certain especially problematic versions of PP that ban every option. Christian Munthe suggests just this constraint, stating that any acceptable interpretation of PP "must not systematically produce decisional paralysis" (2011, p. 36), where a rule produces decisional paralysis when it prohibits all possible actions. Plainly, a version of PP that had this feature could not consistently recommend anything. Hence, Munthe's rule against versions of PP that produce decisional paralysis is a logical consequence of consistency.

But the converse does not hold, that is, consistency does more than merely eliminate versions of PP that ban everything. That is because a version of PP might consistently recommend some precautions but not others. This possibility was discussed in section 2.4.1, where it was suggested that version 2 could consistently justify mitigation measures such as a carbon tax but not a complete phase-out of fossil fuels within a 30-day period. The point here is that whether a version of PP can consistently recommend a specific precaution (e.g., the 30-day phase-out of fossil fuels) *depends not only on the version of PP but also on the details of the precaution being considered.* Merely prohibiting versions of PP that systematically produce decisional paralysis will miss violations of consistency that result from particular features of the precaution. And just as there are some bad versions of PP – those that ban all possible options – there are also bad precautions – those that cannot be consistently justified by any version of PP. For instance, consider preemptive war. Given the obviously severe and difficult-to-contain consequences of warfare, versions of PP capable of recommending a preemptive war as a precaution against some supposed threat would normally also recommend against it.[14] If no version of PP can be found that consistently recommends a proposed precaution, then that is a good indication that other options should be considered. The example of preemptive war is also relevant to the dilemma objection, as Sunstein cites it as an example to support his claim about the incoherence of PP (2005, pp. 4, 60). Thus, overlooking inconsistencies generated by problematic precautions would undermine the attempt to effectively answer the second horn of the dilemma.

### 2.5.2 *Answering the "impossibility theorem"*

In a paper titled "The Precautionary Principle Is Incoherent," Martin Peterson claims to prove that "no version of the precautionary principle can be reasonably applied to decisions that may lead to fatal outcomes" (2006, p. 595). In this section, I argue that Peterson's theorem does not in fact succeed in establishing this conclusion.

According to Peterson, PP applies to "decision making based on qualitative information" (2006, p. 596), which is defined as decisions with a complete ranking of relative probabilities and utilities of outcomes but no quantitative measures of them. Thus, for any two outcomes, one can say whether one is more probable than the other or whether they are equally probable, and the values of any two outcomes can be similarly compared. But we cannot say *how much* more probable or *how much* better one outcome is than another. Peterson (2006) proves two impossibility theorems concerning PP. The first of these concerns a formulation of PP labeled PP$\alpha$, which requires among other things that two actions are equally preferred when they are equally likely to lead to the fatal outcome (2006, p. 597). In what follows, I will assume that the "fatal outcome" corresponds to the harm condition in the version of PP being applied. Peterson's PP$\alpha$, then, conflicts with PP as interpreted here, which can allow for discriminations

[14] See Steel (2013c) for a discussion of this point in relation to the 2003 US invasion of Iraq. Of course, it is possible to imagine fictional situations in which a preemptive war could be justified by PP. The claim here is that such justification is rarely possible in actuality.

among actions with an equal chance of leading to catastrophe. For example, suppose that for acts X and Y the chance of the fatal outcome is ranked as *merely possible*, while for all alternative actions the chance of the fatal outcome is ranked as *scientifically plausible*. Then X and Y are preferred to other actions but not necessarily equally preferred, since efficiency may provide a basis for selecting one over the other. Peterson's second theorem concerns three other formulations of PP, ordered from logically stronger to weaker.[15] The first of these is as follows.

> PP($\beta$): If one act is more likely to give rise to a fatal outcome than another, then the latter should be preferred to the former, given that both fatal outcomes are equally undesirable. (2006, p. 599)

To see how PP($\beta$) connects to the ideas discussed here, recall version 2 of PP, which asserts that if a scientifically plausible mechanism exists whereby an activity can lead to a catastrophe, then that activity should be phased out or significantly restricted. Suppose that this version of PP is being applied to a decision between two options, X and Y. Imagine that it is possible that X would result in a catastrophe, but that there is no scientifically plausible mechanism whereby this could happen. On the other hand, there is a scientifically plausible mechanism whereby action Y leads to catastrophe. In this situation, version 2 consistently recommends X over Y, and hence prefers X to Y, just as PP($\beta$) states. So, in this situation, PP($\beta$) is a consequence of PP as interpreted here. The last two forms of PP considered by Peterson are weakened variants of PP($\beta$), which are qualified by things not included in the interpretation of PP I defend (e.g., a *de minimis* condition). Thus, I consider Peterson's theorem 2 only in relation to PP($\beta$).

That theorem consists in proving that the conjunction of PP($\beta$) and the following three propositions results in a contradiction (2006, p. 601):

*Dominance (D)*: For any actions X and Y, if X produces a result at least as good as Y's in all states and strictly better in some, then X is strictly preferred to Y.

*Archimedes (A)*: "If the relative likelihood of a nonfatal outcome is increased in relation to a strictly better nonfatal outcome, then there is some (nonnegligible) decrease of the relative likelihood of a fatal outcome that counterbalances this precisely" (2006, p. 599).

[15] Sprenger (2012, p. 883) mistakenly presents the conditions of Peterson's theorem 1 as the conditions for his theorem 2. This mistake is significant, as theorem 1 relies on a weaker and more plausible assumption than the Archimedes principle that drives theorem 2.

*Totality (TO)*: The preference ranking is complete, and preference is asymmetric (i.e., if X is preferred over Y, then Y is not preferred over X) and transitive (i.e., if X is preferred over Y, and Y preferred over Z, then X is preferred over Z).

Of these three conditions, D seems unobjectionable. The complete preference ranking in TO might be questioned. However, an examination of Peterson's proof of his theorem 2 shows that the complete ranking assumption in TO is in fact unnecessary. The only components of TO invoked in the proof are asymmetry and transitivity of preference, which seem entirely reasonable as constraints on a decision rule. The big question mark in theorem 2, then, is the Archimedes principle, A. I argue that while A is sensible for decision making involving numerical probabilities and utilities, it is not plausible for reasoning with qualitative information.

To see the intuition behind A, consider a bet with four equally probable outcomes: gain \$10, gain \$5, lose \$5, and lose \$10. Suppose the relative probability of the best outcome, gain \$10, is increased relative to that of the second best, gain \$5. For concreteness, say that the new probabilities are 0.35 for gain \$10 and 0.15 for gain \$5. Obviously, the expected cash payout of the resulting bet is more favorable than the original one. However, it is possible to exactly counteract this positive change by increasing the probability of the worst outcome relative to the second worst, that is, by similarly increasing the chance of lose \$10 with respect to lose \$5. The Archimedes principle works in this case because there is a quantitative measure of utility (namely dollars) and probabilities can be any real number between 0 and 1. Hence, it is possible to find just the right increase in probability of a bad outcome to exactly compensate for an increase in probability of a good outcome.

But it is difficult to see any plausible basis for A in reasoning with qualitative information. As a rationale for A, Peterson writes, "advocates of the precautionary principle must be willing to admit that, to some extent, both the likelihood and the desirability of an outcome matter" (2006, p. 599). However, it is possible to accommodate this observation without being committed to A. In applications of PP, both the harm and knowledge condition are relevant, so likelihood and desirability both matter. In addition, the likelihood of outcomes other than the fatal one matter when selecting among those precautions that can be consistently recommended on the basis of efficiency. The interpretation of PP provided here, then, is not committed to A but nevertheless is compatible with the rationale for A provided by Peterson.

To see why A is problematic for reasoning with qualitative information, consider a case of qualitative reasoning involving an action with four possible outcomes: excellent, good, poor, and catastrophe. In this case, catastrophe would be the "fatal outcome" alluded to in A. Suppose, moreover, that these outcomes are ranked in terms of credibility: respectively, as strong evidence, some evidence, minimal evidence, and merely possible. Now consider another action that increases the credibility of the best outcome relative to the second best: for instance, the excellent outcome is upgraded to very strong evidence while good is downgraded to minimal evidence. This new action is preferable to the original one, since the only change is to increase the likelihood of the best outcome relative to the second best. The Archimedes principle, then, tells us that it is possible to exactly compensate this advantage by increasing the credibility of the worst outcome relative to the second worst. But this is where the difficulties inherent in applying A to qualitative reasoning become apparent.

For what increase of credibility of catastrophe relative to poor would *precisely* offset the advantage accruing from the increase of the credibility of excellent relative to good? I submit that there is no non-arbitrary way to answer such a question. For to know how to answer it, one would need some quantitative measure of the increases and decreases in credibility and utility involved. Is the utility of the excellent outcome twice that of the good outcome or three times? Similarly, how much worse is a catastrophic outcome than a poor one? And what is the exact degree of difference between "very strong evidence" and "strong evidence" and so on? In short, A is sensible only given quantitative measures of credibility and utility that are, by definition, absent in reasoning with qualtitative information. To insist that a theory of decision with qualitative information be committed to A is to unreasonably demand that such theories draw arbitrary distinctions.

## 2.6 Conclusions

Given its prominent role in disputes concerning a variety of pressing contemporary environmental issues, efforts to clarify PP and its logical implications are especially urgent (Gardiner 2011, pp. 411–14). This chapter has aimed to take one step in this direction by answering the dilemma objection to PP, according to which PP is – depending on whether it is given a "weak" or "strong" interpretation – either vacuous or irrational. I argued that the distinction between "weak" and "strong" PP is misleading and should be replaced with a more accurate contrast between PP as a meta-rule and a decision rule. Meta versions of PP do not select among distinct environmental policies but instead require that the decision-making procedures used for these purposes should not be susceptible to paralysis by scientific uncertainty. Such claims are substantive because they often will recommend against basing environmental policy decisions on cost–benefit analysis. To answer the second horn of the dilemma, I developed a new explication of proportionality in terms of the twin theses of consistency and efficiency.

CHAPTER 3

# *The unity of the precautionary principle*

## 3.1 *The* precautionary principle?

Advocates and detractors alike often claim that it is a mistake to refer to "*the* precautionary principle," suggesting that it is in fact many different things rather than a unified doctrine. For instance, in the words of Andrew Jordan and Timothy O'Riordan,

> [PP] is neither a well-defined nor a stable concept. Rather, it has become the repository for a jumble of adventurous beliefs that challenge the status quo of political power, ideology, and environmental rights. Neither concept has much coherence other than it is captured by the spirit that is challenging the authority of science, the hegemony of cost–benefit analysis, the powerlessness of victims of environmental abuse, and the unimplemented ethics of intrinsic natural rights and intergenerational equity. (1999, p. 16)

Although harshly attacking rather than advocating PP, Gary Marchant and Kenneth Mossman make similar claims about its lack of internal coherence:

> The precautionary principle is different, however, with respect to both the extent of its ambiguity and the imperviousness of that ambiguity to resolution. While other risk decision-making concepts can be, and indeed have been, subject to more precise refinements, the precautionary principle not only has not been further clarified but, by its very nature, *cannot* be made more precise.[1] (2004, p. 13, italics in original)

[1] To support these claims, Marchant and Mossman (a) describe differing interpretations of PP (2004, pp. 9–13), (b) state a version of the dilemma objection (2004, p. 14), and then (c) cite Jordan and O'Riordan (1999). But disagreements about how to interpret or apply PP are hardly surprising and do not distinguish it from other decision-making approaches such as cost–benefit analysis, a point illustrated by the discussion of disputes about future discounting in Chapter 6. Nor does (a) show that it is impossible to clarify PP or to make progress on resolving disagreements about its interpretation. Regarding (b), see Chapter 2 for a critique of the dilemma objection. As for (c), there is good reason not to take Jordan and O'Riordan's statement as authoritative, as this chapter shows (also see Gardiner 2006, pp. 39–40).

To support such assertions, one might cite different interpretations of PP that have been launched from seemingly divergent theoretical perspectives, including the maximin rule (Ackerman 2008a; Gardiner 2006; Hansson 1997), catastrophe principles (Hartzell-Nichols 2012; Sunstein 2005, chapter 5), minimax regret (Chisholm and Clarke 1993), robust adaptive planning (Doyen and Pereau 2009; Johnson 2012; Mitchell 2009; Sprenger 2012), and alternatives assessment (O'Brien 2000; Tickner 1999; Tickner and Geiser 2004).

This multitude of outlooks on PP raises some difficult questions. Do they show that PP really is no more than a jumble of loosely connected, sometimes contradictory, vaguely pro-environmentalist proposals, as the two passages quoted above assert? If so, it would be difficult to understand how PP could be an effective guide for decisions on environmental policy, or any other topic. Or is there some more general interpretation that provides a reasoned basis for identifying where these several proposals coincide with PP and where they depart from it? In this chapter, I argue that the interpretation of PP advanced here supports a positive answer to this latter question.

I begin by examining arguments for the disunity of PP, and I make the case that none provide good reason to accept their conclusion. Although statements of the disunity of PP are fairly common, arguments for such claims are relatively rare. I focus on the most detailed argument of this kind I know, which is due to Mariam Thalos (2009, 2012). Thalos points to a diverse array of ideas related to precaution and insists that no single perspective on PP could capture all of them. Her argument is framed as a critique of Gardiner's proposal that a version of the maximin rule restricted by conditions inspired by John Rawls's famous discussion of the "original position" (1971) constitutes a "core" of PP. Although I agree that Gardiner's attempt to bring some unity to PP is problematic, I argue that Thalos fails to establish her more general conclusion that PP is inherently diffuse and fragmented. I examine the various aspects of precaution Thalos describes, and for each one I explain how the interpretation I propose incorporates it or why there are good reasons for not making it part of PP.

From here I proceed to show how the interpretation I propose clarifies the relationship between PP and the several approaches canvassed above – catastrophe principles, maximin, minimax regret, robust adaptive planning, and alternatives assessment. In sections 3.3.1 and 3.3.2, respectively, I show how the catastrophe precautionary principle advocated by Hartzell-Nichols (2012) and the Rawlsian version of maximin advocated by Gardiner (2006, 2010) are special cases of PP as I interpret it. Next, in section 3.3.3, I consider a proposal to interpret PP in terms of minimax regret (Chisholm and Clarke 1993). I explain how minimax regret agrees with maximin under the Rawlsian conditions considered by Gardiner. However, advocates of minimax regret argue that the maximin rule is not precautionary because it is incapable of justifying precautions whenever it is uncertain whether a catastrophe can be prevented (Chisholm and Clarke 1993, pp. 113–14). I explain how MPP supports this argument and hence that PP and maximin, despite agreeing in the Rawlsian conditions discussed by Gardiner (2006), are not equivalent in general. But I also argue that PP, which is intended to be applicable to decision making with qualitative information, is not the same as minimax regret, which presumes a quantitative utility measure.

Finally, in section 3.4, I turn to robust adaptive planning and alternatives assessment. In section 3.4.1, I explain how robust adaptive planning and alternatives assessment are relevant to PP as a means for finding precautions that can be consistently recommended and for enhancing the efficiency of proposed precautions. In section 3.4.2, I examine disagreements about whether robust adaptive planning and PP are conflicting or mutually supporting ideas and argue in favor of the latter option.

## 3.2 Arguments for disunity

In this section, I examine arguments for the unavoidable disunity of PP, and I make the case that none of these arguments withstands scrutiny. Perhaps the most common argument of this kind simply consists of describing the large number of distinct and sometimes disagreeing formulations of PP (Hartzell-Nichols 2013; Marchant and Mossman 2004). But such an argument is not very compelling, as distinct forms of PP do not rule out the possibility of substantive uniting themes. Nor does a diversity of formulations of PP preclude giving good reasons for preferring one approach rather than another in cases of genuine disagreement.

Consequently, I focus on a more sophisticated argument, due to Thalos (2009, 2012), that rests on identifying disparate and possibly conflicting philosophical rationales for precaution. My approach will be to list the aspects of precaution suggested by Thalos and then to explain either how they are incorporated within the interpretation of PP proposed here or why there is good reason to reject them as elements of PP.

1. Thalos argues that, although it is common to restrict PP to cases involving uncertainty about the probabilities of outcomes, the concept of precaution can be relevant to cases in which probabilities are known (2009, p. 43).

2. Thalos notes that the "pre" in "precaution" could refer to (a) not waiting for conclusive evidence before acting; (b) lexically ordering values (e.g., treating environmental harms as always more weighty than economic harms); (c) taking action before the harm is manifested; or (d) the idea that it is better to prevent an injustice before it happens than to attempt to rectify it after the fact (2009, pp. 44–5).
3. Thalos points out that the concept of precaution may be relevant even when the course of a policy cannot be modified once it is set in motion (2009, p. 47).

Regarding these various aspects of precaution, Thalos writes, "it is quite unreasonable to anticipate that there will be a 'core' thought or principle that captures all of their different concerns, even where one fully acknowledges these as worthy principles, goals and concerns" (2009, p. 45).

One might respond to Thalos's argument by pointing out that its target is not PP but a vaguely understood notion of precaution. It would be unreasonable to insist that PP must capture every idea that might be associated with the concept of precaution generally. After all, some of those ideas might be misguided and some may be jointly inconsistent. But even putting this issue aside, it is questionable whether Thalos's aspects of precaution are really so diverse and unrelated as she suggests. For example, consider (a), (c), and (d) from entry 2: (a) is typically a necessary condition for accomplishing (c), and (d) is a moral argument for (c) that is relevant when the harm in question involves an injustice. Far from being disparate, then, (a), (c), and (d) are closely interconnected. Moreover, all of these elements are readily accommodated in the interpretation of PP proposed here. Thus, (a) is a consequence of MPP, and taking precautions before the damage is done (i.e., (c)) is what MPP aims to facilitate. In addition, irreversible injustices, as discussed in (d), would be a factor accentuating the severity of harm and hence are relevant to the harm condition in an application of PP.

Next, consider item 1, which states that the concept of precaution may be relevant even when probabilities are known. This claim is also in agreement with the interpretation of PP proposed here, which rejects the standard decision-theoretic definition of uncertainty as ignorance of probabilities (see Chapter 5). According to the proposal advanced here, scientific uncertainty is defined as the absence of a model whose predictive validity for the task in question is empirically well confirmed. Scientific uncertainty in this sense might or might not be quantifiable by probabilities. In addition, PP could be relevant even in an ideal case in which scientific uncertainty is entirely absent. That is because some types of value judgments are especially problematic from the perspective of PP, most notably, a value judgment that the future is worth relatively little in comparison to the present. Chapter 6, which discusses the controversial topic of future discounting, explains why PP should be committed to intergenerational impartiality.

In item 3, Thalos presumably means to argue against the idea that precaution should be understood in terms of adaptive management, which recommends policies that are designed to be responsive to incoming information. Again, I agree with Thalos that PP can be relevant even when adaptive management approaches are not feasible. But this is also encompassed by the interpretation of PP proposed here. As discussed below in section 3.4, adaptive management is a useful means for enhancing efficiency in many, if not most, environmental policy decisions. But the three core themes of PP do not presuppose adaptive management in any conceptual sense and hence remain applicable to cases in which adaptive management is not possible.

Finally, consider (b) from entry (2), according to which values are lexically ordered, with some always taking precedence over others. In relation to PP, an example of this idea would be that environmental harms always trump adverse economic impacts, so any environmental gain, no matter how minor, is always worth incurring any economic cost (Thalos 2012, p. 172). My response here is simply that this idea should not be a component of PP because it leads to an absolutist approach that is incompatible with proportionality. Such a lexical ordering is also problematic because injustices that can accentuate the severity of environmental harms, such as climate change, are often inextricably tied to economic harms. For example, consider salinization of soil in low-lying coastal areas of Bangladesh due to the rising sea level. This is a classic example of an injustice resulting from climate change, yet it is also an inherently economic harm since it entails the impoverishment of Bangladeshi farmers.

However, I share Thalos's skepticism about Gardiner's (2006) proposal that a restricted version of the maximin rule can function as a useful "core" of PP. But I argue that the shortcomings of Gardiner's proposal are due to specific details of his approach, rather than to an alleged inevitable disunity of PP. The primary difficulty with Gardiner's approach is that the restricted circumstances he considers result in a relatively easy decision, with the result that several normally diverging decision rules yield the same result as maximin. Consequently, Gardiner's proposal leaves it unclear which rule to adhere to when these conditions are relaxed. Gardiner's maximin proposal is discussed in further detail below in section 3.3.2.

## 3.3 Catastrophe, maximin, and minimax regret

This section examines three principles proposed as partial explications of PP. The first of these is the catastrophic precautionary principle proposed by Hartzell-Nichols (Hartzell 2009, Hartzell-Nichols 2012). The second is a restricted version of the maximin rule suggested as an interpretation of PP by several authors (Ackerman 2008a; Ackerman and Heinzerling 2004; Gardiner 2006). I show that the first two are special cases of PP as interpreted here. Moreover, I explain how the greater generality afforded by my account of PP is important for applications to cases such as climate change and toxic chemicals. Finally, I consider the relationship between the maximin rule and minimax regret. I explain how MPP supports an argument made by proponents of minimax regret charging that the maximin rule fails to be precautionary when it is uncertain whether a catastrophic outcome can be prevented (Chisholm and Clarke 1993). Nevertheless, I explain why PP as I interpret it is distinct from minimax regret.

### *3.3.1 The catastrophic precautionary principle*

Hartzell-Nichols' catastrophic precautionary principle (CPP) is as follows.

> *Catastrophic Precautionary Principle (CPP)*
> *Appropriate precautionary measures should be taken against threats of catastrophe*
> Where:
>
> - Threats of catastrophe are those in which many millions of people could suffer severely harmful outcomes (defined as severely detrimental to human health, livelihood, or existence).
> - A precise probability of a threat of harm is not needed to warrant taking precautionary measures so long as the mechanism by which the threat would be realized is well understood and the conditions for the function of the mechanism are accumulating.
> - Appropriate precautionary measures must not create further threats of catastrophe and must aim to prevent the potential catastrophe in question.
> - Imminent threats of catastrophe require immediate precautionary action.
> - Threats of catastrophe that involve an imminent threshold or point of no return for effective precautionary action (beyond which precautionary measures are limited or unavailable) also require immediate precautionary action aimed at preventing this threshold from being crossed.
> - Non-imminent threats of catastrophe might warrant further study before further precautionary measures are implemented, provided a delay in taking precautionary measures will not prevent such measures from effectively preventing the catastrophic outcome in question.

- The likelihood (or probability) of a catastrophically harmful outcome may affect *what* precautionary measures are taken, but not *that* precautionary measures should be taken. That is, a low probability outcome/event might warrant more minimal precautionary measures than a similar high probability outcome/event (which might require aggressive mitigation measures). (Hartzell-Nichols 2012, pp. 160–1, italics in original)

Consider, then, the relationship between CPP and the three core themes of PP articulated here. The second bullet in CPP is intended to meet the demand of MPP that decision rules used for environmental policy should not be immobilized by scientific uncertainty. Likewise, the elements of the "tripod" – harm condition, knowledge condition, and recommended precaution – are clearly present. The specification of the tripod in CPP is very similar to version 2 of PP discussed in the preceding chapter, which asserted, "If a scientifically plausible mechanism exists whereby an activity can lead to a catastrophe, then that activity should be phased out or significantly restricted." The connection between CPP and proportionality, on the other hand, is less obvious and more interesting.

The third clause in CPP requires that the precaution itself not generate a catastrophe. This ensures that consistency – one of the two elements of proportionality – is satisfied. Recall that consistency demands that the version of PP used to justify the precaution should not also recommend against carrying out that very precaution. In this case, CPP is the version of PP in question. Thus, if it is known that the recommended precaution will not lead to a catastrophe, then the precaution is consistent with respect to CPP. The next two bullets in CPP, which state that immediate precautions are required when confronted with imminent threats of catastrophe, are also consequences of PP. This is because consistency excludes ineffectual precautions as well as precautions that generate excessive side effects (see section 2.4.2). For suppose that there is a plausible scientific mechanism whereby a catastrophe may occur in the near future. Then any precaution that does not take prompt action will be ineffective, and hence version 2 will recommend against such "precautions" for the same reason it recommends against sitting idly by and doing nothing.

The final two bullets of CPP are also straightforward consequences of proportionality explicated in terms of consistency and efficiency. The second-to-last bullet states that less rapid precautions can be appropriate when the potential catastrophe is not imminent. That is certainly compatible with consistency, and if a more measured precaution is more efficient, then it would be required by proportionality. The final bullet reflects a central feature of proportionality, namely that the harm and knowledge conditions satisfied in the case influence the choice of precaution. In addition, PP as defended here agrees with Hartzell-Nichols' emphasis that a version of PP that can be consistently applied to the case at hand is sufficient to establish a duty to act even though some details of the precaution may remain to be ironed out. Hartzell-Nichols' CPP, then, is a proposition that anyone who agrees with the three core themes of PP should accept.

However, CPP is also clearly less general than those three themes. Most apparently, the harm and knowledge conditions in Hartzell's CPP are not adjustable but are fixed at "catastrophe" and "the mechanism by which the threat would be realized is well understood and the conditions for the function of the mechanism are accumulating," respectively. The restriction to catastrophes seems motivated in part by a desire to avoid the incoherence objection discussed in Chapter 2 (see Hartzell 2009, p. 73). But since consistency is what blocks the incoherence objection, setting the harm and knowledge conditions to pre-specified fixed points unnecessarily restricts the range of application of PP. While CPP applies only to catastrophes, PP is commonly thought to have a much broader range of application – for instance, to toxic chemicals and genetically modified foods.

The second major limitation of CPP is that consistency is guaranteed by a requirement that is excessively strict. The formulation of consistency in CPP states that "precautionary measures must not create further threats of catastrophe." The difficulty here is that the knowledge condition required to establish consistency by this rule appears to be unattainable in many circumstances in which PP would be applied. On the natural reading, showing *no threat* of catastrophic side effects means showing that such effects are *impossible*. This reading is reinforced by Hartzell-Nichols' clarification that "a precautionary measure cannot introduce a new threat of catastrophe, *however remote or unlikely that threat is*" (2012, p. 165, italics added). But such a high standard of proof would have the effect of precluding almost all precautions. For example, consider aggressive climate change mitigation whose economic impacts are uncertain and likely to be very wide ranging.[2] A worldwide economic depression could qualify

[2] Surprisingly, Hartzell-Nichols does not address the issue of consistency in relation to her proposal for aggressive climate change mitigation (see Hartzell 2009, pp. 185–7; Hartzell-Nichols 2012).

as catastrophic given Hartzell-Nichols' standard of severely harmful outcomes suffered by many millions of people. And although there may be good scientific reasons to believe that such side effects would not occur as a result of climate change mitigation, it is doubtful that anyone is in a position to definitively rule them out of the sphere of possibility altogether. Thus, scientific uncertainty about the effects of precautions makes it difficult for Hartzell-Nichols' CPP to justify substantial action. Consequently, although CPP is a special case of PP as interpreted here, it is a special case of very narrowly limited scope. It fails to recognize that MPPs' demand that scientific uncertainty not be a reason for inaction also pertains to uncertainty about effects of precautions.[3]

Recall how PP as construed here addresses this issue. The threshold of scientific plausibility in an application of consistency is determined by the knowledge condition in the version of PP used to justify the precaution. In the case of climate change, for instance, that knowledge condition might be a scientifically plausible mechanism. Hence, it is not necessary to show that catastrophic effects of aggressive mitigation measures are impossible; it is only necessary to show that there is no known scientifically plausible mechanism by which they would occur. However, Hartzell-Nichols explicitly rejects this approach, insisting that CPP does not permit precautions that introduce any possibility of catastrophe whatever (2012, p. 166). In defense of this position, Hartzell-Nichols writes:

> If all a Precautionary Principle required of us is that we take precautionary measures that reduce a threat of harm or even a threat of catastrophe, we may be justified in taking superficial and ineffective measures that we know to raise their own threats of catastrophe when there are other precautionary measures available that do not invite such threats. (Hartzell-Nichols 2012, p. 166)

But ineffectual "precautions" with known yet avoidable harmful side effects are not compatible with proportionality. Consistency places a strong emphasis on effective precautions, as explained above. And if equally or more effective precautions with less harmful side effects are available, then efficiency demands that they be chosen.

[3] The points made regarding Hartzell's CPP in this section would also apply similarly to Sunstein's Anti-Catastrophe Principle (2005, pp. 114–15). Like Hartzell's CPP, Sunstein's Anti-Catastrophe Principle is unnecessarily restricted to catastrophes and incorporates a version of consistency that appears excessively strict.

### *3.3.2 Maximin*

The maximin rule recommends that one choose the action with the least bad worst-case outcome – in other words, that one act so as to maximize the minimum that one is assured of attaining. Several authors have been struck by the analogy between the maximin rule and PP (Ackerman 2008a, pp. 87–8; Ackerman and Heinzerling 2004, pp. 225–6; Gardiner 2006; Hansson 1997). Both restrict the set of factors that are relevant to the decision so as to focus attention on adverse outcomes, and both recommend taking precautions in the face of uncertain threats, where "uncertainty" may involve ignorance of probabilities of relevant outcomes. In the remainder of this section, I examine a restricted version of maximin inspired by the writings of John Rawls (1999).

Few authors defend the maximin rule without qualification.[4] Typically, advocates propose that it be followed only under rather narrowly defined circumstances. The most central restriction is what can be called pure uncertainty:[5] a total lack of information about the probabilities of the relevant possible scenarios. In the case of climate change, for instance, this could mean that we have no idea how likely it is that sea levels will rise one-half or one or two or three meters by the end of the century. However, justifications of maximin typically assume that the exhaustive list of possible future scenarios is known with certainty. It is also assumed that, for any action and any scenario, the consequences of performing that action when that scenario obtains are known.

Frank Ackerman (2008a, pp. 233–5) cites a little known theorem published by Kenneth Arrow and Leonid Hurwicz, which shows that under conditions of pure uncertainty, and given a few plausible logical assumptions about optimality, the optimal action depends solely on the best-case and worst-case outcomes (Arrow and Hurwicz 1972). In other words, given the conditions assumed in the theorem, intermediate results of actions are irrelevant for deciding what to do, and only their extreme maximum and minimum consequences matter. To get from here to the maximin rule, some additional reason is needed to focus only on the worst-case outcomes. Stephen Gardiner (2006, p. 47) suggests two conditions for this purpose, which he adopts from Rawls (1999). According to the first of these conditions, "the decision-makers care relatively little for potential gains that

[4] Shrader-Frechette (1991, chapter 8) is one of the few who do, although she does not associate maximin with PP.

[5] I borrow this term from Ackerman (2008a, p. 87).

might be made above the minimum that can be guaranteed by the maximin approach" (1999, p. 134). According to the second, "Rejected alternatives have outcomes that one can hardly accept" (ibid.).

Let S be a set of possible scenarios each of which satisfies some standard of scientific plausibility. Then, Gardiner's proposal amounts to a recommendation to take a precaution when the following conditions are satisfied:

(a) No information exists upon which to judge the relative probability of members of S (pure uncertainty).
(b) There is no scenario in S wherein catastrophe occurs when the precaution is taken.
(c) For every scenario in S, the precaution achieves results that are close to the best that could have been achieved in that scenario.
(d) Every alternative to the precaution leads to catastrophe in at least one scenario in S.

For convenience, I will collectively label (a) through (d) the *Rawlsian conditions*. Condition (a) is pure uncertainty, while conditions (b) and (c) are teased apart from Rawls's requirement that decision makers care very little for gains above what is assured by following the maximin rule. That claim entails that the precaution guarantees an acceptable minimum (and hence assures that catastrophe is avoided). And since decision makers "care very little" for potential gains above that minimum, the results of the precaution are always close in value to the best that could have been achieved by any action, which is to say the cost of the precaution is always very low. Notice that (b) and (c) are logically independent. The precaution might be assured of preventing catastrophe, while imposing significant unnecessary costs in some scenarios (i.e., (b) could be true yet (c) false). Conversely, even if the precaution is assured of being very low cost, there might be some scenarios in which catastrophe cannot be averted by any action (i.e., (c) could be true but (b) false). Finally, condition (d) is an expression of the final Rawlsian condition that every alternative to the precaution is associated with "grave risks."

Several critics of Gardiner's proposal have argued that one or more of the Rawlsian conditions are unlikely to obtain in actual applications of PP. For example, Sunstein argues that environmental regulations are rarely assured of being low cost in all scenarios (2005, p. 112). Gardiner responds to this objection by defending condition (c) with respect to climate change mitigation, suggesting that the "care little for gains" requirement could be interpreted to include cases of costs that are relatively large in dollar terms (2011, pp. 413–14). This reply is rather unpersuasive, since citizens and the politicians they elect generally do seem to care a great deal about small changes in GDP. But even if condition (c) could be rescued in the case of climate change mitigation, condition (b) would remain problematic. Condition (b) will fail whenever uncertainty exists about whether the precaution will prevent catastrophe (see Chisholm and Clarke 1993, pp. 113–14). Asserting condition (b) with respect to climate change requires definite knowledge that current atmospheric $CO_2$ concentrations do not already make some catastrophic effects of climate change a foregone conclusion, yet it is doubtful whether climate science can provide complete assurances that such "points of no return" have not already been crossed.

In addition, condition (a) is very restrictive and unlikely to obtain in many circumstances to which PP is normally thought relevant. In many environmental policy problems, some information about probabilities exists, a point illustrated by climate change. Assessment reports of the IPCC have included rough consensus estimates of the probabilities of a number of potential impacts for over a decade, including rises in sea levels. So, the situation in this case and many others is not that *nothing* is known about probabilities of various possible scenarios, but rather that knowledge of these probabilities is incomplete in crucial ways.

So, one objection to the Rawlsian conditions is that they are unlikely to be satisfied with respect to many environmental issues. A different complaint is that these conditions, even if granted, fail to support an argument for maximin over alternative decision rules. Notice that conditions (b) and (d) alone are sufficient for the maximin rule to recommend the precaution. Given these two conditions, the worst-case outcome of the precaution is clearly superior to the worst-case outcome of every other action. Hence, conditions (a) and (c) merely make it possible that *other* decision rules, some of which may not be consistent with maximin in general, *also* recommend the precaution in this case. For example, Greg Bognar (2011) emphasizes this point in a critique of Gardiner's approach, observing that maximizing expected utility would also lead to enacting the precaution when the Rawlsian conditions obtain, provided that a uniform prior probability distribution should be used in cases of pure uncertainty. In the next section, I explain why minimax regret also accords with the maximin rule under the Rawlsian conditions. Moreover, the interpretation of PP advanced here also recommends taking the precaution when the Rawlsian conditions obtain, since the precaution in that circumstance is the only action that could be consistently recommended by PP.[6]

[6] See section 2 of the Appendix for a more detailed discussion of this claim.

From the perspective of the interpretation of PP I propose, the Rawlsian conditions are a special case of a more general sufficient condition for recommending a precaution. A precaution is recommended by PP as interpreted here if:

(e) there is an adequate version of PP that consistently recommends the precaution;

(f) no adequate version of PP consistently recommends any action other than the precaution.

Inadequate versions of PP are those eliminated on the grounds that they are either paralyzing, in the sense of prohibiting everything, or uninformative, in the sense of allowing everything.[7] Conditions (e) and (f), while sufficient, are not necessary for PP to justify an action. For instance, there might be several precautions that can be consistently recommended by an adequate version of PP, yet one precaution may nevertheless be recommended over the others on grounds of efficiency. Nevertheless, conditions (e) and (f) avoid the problematic assumptions appealed to in Gardiner's defense of the maximin rule. Pure uncertainty is not required. Nor is it required that the precaution is low cost in every scientifically plausible scenario. Finally, it is not required that there is no scenario in which catastrophe occurs if the precautionis performed.

Several alternative decision rules to maximin, therefore, coincide with it in the narrowly restricted Rawlsian conditions. This raises serious problems for regarding maximin a "core" of PP, because Gardiner's proposal does not provide grounds for claiming that PP should be understood in terms of the maximin rule rather than one of these alternatives. A more adequate approach to providing a unifying interpretation of PP, then, should not be tied to the Rawlsian conditions, and it should be capable of providing a reasoned basis for selecting between differing proposed explications of PP. I claim that the interpretation of PP proposed here does both of these things.

### *3.3.3 Minimax regret*

The concept of minimax regret was first developed by economists Graham Loomes and Robert Sugden (1982) to explain divergences between actual human decisions and what would be expected according to certain models

[7] See section 1.3 of the Appendix for a formal definition of acceptable versions of PP. See section 2 of the Appendix for a more detailed explanation of the more general sufficient condition.

Table 3.1 *Costs*

| | $s_1$ | $s_2$ | $s_3$ |
|---|---|---|---|
| Precaution | $e$ | $e$ | $e + c$ |
| No precaution | $c$ | $m$ | $c$ |

of rational choice commonly assumed in economics. However, minimax regret is of interest here as a normative rule for guiding decisions in circumstances of scientific uncertainty. Minimax regret has been advanced as an explication of PP (Chisholm and Clarke 1993), and regret is a fundamental concept in some approaches to robust adaptive planning, as will be discussed below. In this section, I consider the relationship among the maximin rule, minimax regret, and PP. I describe minimax regret and show how it agrees with the maximin rule when the Rawlsian conditions obtain. Next, I explain why MPP supports an argument for preferring minimax regret to the maximin rule when it is uncertain whether catastrophe can be avoided. However, PP as interpreted here is an approach to decision with qualitative rankings of harm and knowledge conditions, and hence is distinct from minimax regret, which presumes a quantitative utility measure.

The concept of minimax regret is best explained by means of an example. In Table 3.1, rows represent possible actions and columns possible states of the world. The precaution is intended to prevent a catastrophe, whose disutility is represented by $c$, but implementing the precaution involves an expense, $e$. In the first state, $s_1$, catastrophe occurs if the precaution is not taken but will be averted if it is. In $s_2$, catastrophe does not occur no matter which option is chosen. In this state, some harm, whose cost is denoted by $m$, less severe than a catastrophe occurs if the precaution is not taken. Finally, in $s_3$ the catastrophe is unpreventable and occurs no matter which action is taken. Each cell in the table gives the costs incurred for that combination of action and state of the world. Thus, in $s_1$, the precaution results in the expense $e$, while not taking the precaution results in catastrophe, $c$, while in $s_3$ the precaution results in $e + c$ and not taking the precaution in $c$. For convenience, I treat $c$, $e$, and $m$ as absolute values of costs, so, for instance, $c > e$ if the cost of the catastrophe is greater than the cost of implementing the precaution.

For any combination of action and state, regret is defined as the difference between the result of that action in that state and the best result that could

Table 3.2 *Regrets*

| | $s_1$ | $s_2$ | $s_3$ |
|---|---|---|---|
| Precaution | 0 | $e - m$ | $e$ |
| No precaution | $c - e$ | 0 | 0 |

have been obtained in that state by any action. Thus, supposing that $c > e > m \geq 0$, Table 3.1 leads to the regrets specified in Table 3.2. For example, in $s_1$ the precaution is the action that yields the best result, and hence its regret is zero, while the regret of not taking the precaution is $c - e$. The minimax regret rule, then, recommends that one select the action that has the smallest maximum regret. The maximum regret for the precaution in this case is $e$, while the maximum regret for not taking the precaution is $c - e$. If the expense of the precaution is less than half the cost of a catastrophe (i.e., if $c > 2e$), then minimax regret recommends that the precaution be taken.

Consider, then, the relationship between the maximin rule and minimax regret. First, notice that the two coincide under the Rawlsian conditions. This can be easily appreciated through an examination of Tables 3.1 and 3.2. First, notice that condition (b), according to which enacting the precaution guarantees that catastrophe does not occur, eliminates scenario $s_3$ from both tables. Next, condition (c) ensures the cost of the precaution, $e$, is always tiny in comparison to $c$. Finally, condition (d) says that for every alternative to the precaution, there is some scenario in which it results in catastrophe, as is the case in $s_1$. Thus, given conditions (b), (c), and (d), the precaution has both the least bad worst-case outcome and the smallest maximum regret.

However, the maximin rule and minimax regret can generate contrary recommendations when the Rawlsian conditions do not hold, a point which can be illustrated by reference to Tables 3.1 and 3.2. Letting all three states be possible and assuming as before that $c > 2e$, the maximin rule recommends against the precaution. That is because the worst-case outcome of not taking the precaution, $c$, is less bad than that of the precaution, $c + e$. Yet minimax regret recommends the precaution under these same conditions, as explained above. Anthony Chisholm and Harry Clarke (1993, pp. 113–14) use an example of the same form – with climate change mitigation as the precaution – to argue that minimax regret is preferable to maximin as an interpretation of PP. They point out that the maximin rule is incapable of justifying a precaution, even if it is extremely low cost, whenever it is uncertain whether the precaution will be able to prevent catastrophe. Although they do not present it as such, Chisholm and Clarke's argument is a straightforward application of MPP. In the circumstances in question, the maximin rule allows uncertainty to be a reason for failing to take precautions in the face of serious threats. Therefore, MPP advises against adhering to the maximin rule in those circumstances.

Sven Ove Hansson (1997, pp. 297–9) discusses Chisholm and Clarke's argument, but in a case of one person's modus ponens being another's modus tollens, takes it as proof that minimax regret is not the correct interpretation of PP. However, Hansson's response is forceful only to the extent that there is some independent reason that unequivocally establishes the maximin rule as the definition of PP. Chisholm and Clarke are certainly not the only ones to question this association (see Godard 2000, pp. 26–8; Hartzell 2009, pp. 60–1). So, Hansson bears a burden of argument to show that PP and maximin are indeed one and the same. And he does defend this position as follows: "The maximin rule gives priority to avoiding the worst that can happen. Therefore, it corresponds closely to the intuitive notion of being cautious, or exercising precaution" (1997, p. 295). Yet as an argument that PP must be interpreted in terms of the maximin rule, this leaves much to be desired. First, it is questionable whether the maximin rule always coincides with intuitive ideas about precaution: for instance, consider a person who enjoys riding a motorcycle sans helmet for the joy of feeling the wind in her hair. If $e$ represents the cost of foregoing this pleasure, while $c$ denotes the catastrophic costs of a traumatic brain injury, then one plausibly has an example that exemplifies Tables 3.1 and 3.2 and the assumption that $c > 2e$. Maximin recommends against wearing the helmet on the grounds that the worst case occurs when one is flattened by a tractor-trailer and the helmet fails to prevent severe brain trauma. In that case, one suffers both the catastrophe of serious brain injury and the added cost of foregone pleasure. Yet my intuition, and the intuition of most others, I strongly suspect, is that putting on the helmet is nevertheless a reasonable precaution, while to ride helmetless is to throw caution to the wind. Secondly, even if the maximin rule were the undeniably correct way to understand intuitive ideas about precaution, it would not follow that it is the correct way to interpret PP. After all, PP is not merely an intuitive notion but a concept elaborated in a variety of legal and philosophical contexts. Moreover, the idea that uncertainty should not be a reason for failing to take precautions against serious threats (i.e., MPP) is one of its central elements.

Table 3.3 *Qualitative rankings*

| | $s_1$ | $s_2$ | $s_3$ |
|---|---|---|---|
| Precaution | modest cost | modest cost | catastrophe |
| No precaution | catastrophe | little or no cost | catastrophe |

So PP as interpreted here definitely supports Chisholm and Clarke's argument against the maximin rule in cases (such as the motorcycle-helmet example) that include dire scenarios wherein catastrophe is unpreventable. However, it does not follow from this that PP, as understood here, is equivalent to minimax regret. In fact, there is a simple reason why that cannot be the case. According to the proposal advanced here, PP as a decision rule is applicable to cases involving qualitative rankings, wherein knowledge and harm conditions can be ranked but numerical weights cannot be assigned to them in meaningful ways. Yet minimax regret presumes a quantitative measure of the values of outcomes. For otherwise, what sense could be given to talk of the *difference* (e.g., $c - e$) between two costs?

Consequently, PP as understood here must provide a distinct rationale from minimax regret for recommending that, for instance, you wear a helmet while riding a motorcycle even if you enjoy the flow of wind in your hair. That rationale is related to the fact that the harm condition in versions of PP is a qualitative category, such as catastrophe or irreversible environmental damage. An unavoidable consequence of representing harm conditions with qualitative categories is that outcomes differing in severity of harm in some minor respect will often be lumped into the same category. The qualitative treatment of harm conditions is closely related to MPP. For requiring an exact value of the severity of harm associated with every outcome would result in just the sort of paralysis by analysis that MPP seeks to avoid. Suppose in the motorcycle-helmet example, then, that the harm condition is the catastrophe of severe brain trauma. Since the cost of the foregone pleasure of feeling the wind in your hair is trivial by comparison, serious brain trauma plus not feeling the wind is also categorized as a catastrophe. Table 3.3 is an example of what Table 3.1 might look like if represented using qualitatively ranked harm conditions.

Consider, then, why PP would recommend the precaution in the situation represented in Table 3.3. As explained in section 2.4.1, a possible scenario in which harm occurs counts as a reason against an action only if there is some possible alternative action that could avoid that harm in that scenario. Consequently, dire scenarios in which catastrophe occurs no matter what we do (such as $s_3$) cannot be a reason for preferring one action over another. Given consistency, then, PP can only recommend an action be taken which avoids catastrophe in the remaining two states, $s_1$ and $s_2$, which means that the precaution is the only action that could be consistently recommended in Table 3.3.

This example also suggests a divergence between PP and minimax regret. A key part of what makes minimax regret seem like a more plausible construal of precaution in the motorcycle-helmet example is that the cost of wearing the helmet appears trivial in comparison to the catastrophic harm. The same goes in the case of climate change mitigation: a slight reduction in economic growth seems exceedingly minor in comparison to a catastrophe in which sea level rise forces the widespread abandonment of densely populated, economically and culturally significant coastal areas. In neither case is it plausible to suggest that the added cost puts the overall harm over the threshold into a qualitatively more severe category. Yet if the cost of the precaution is more significant, then the harm the precaution seeks to prevent combined with the cost of the precaution itself might be reasonably judged to be qualitatively worse than the threatened harm alone. And that might be so even if $c > 2e$, that is, even if minimax regret recommends the precaution.

Indeed, an example given by Hansson (1997, pp. 297–8) to argue against minimax regret as an interpretation of PP has precisely this character. In Hansson's example, one considers whether to attempt to save a dying lake by adding iron acetate to it or to do nothing. The iron acetate is assumed to be available free of charge, but it is known that adding it to the lake would result in some harmful effects on land animals that drink from the lake, although it is believed that this harm is less severe that the death of the lake itself. In the example, three states are believed to be possible. In $s_1$, the lake will be saved if iron acetate is added but will die if it is not. In $s_2$, the lake will regenerate if nothing is done, and adding iron acetate will make no difference. Finally in $s_3$, the lake dies no matter what. Hansson supposes that the costs for each act–state combination are as given in Table 3.4. Like the motorcycle-helmet example, this is a case in which maximin recommends against the precaution, while minimax regret recommends in favor. That is, the worst case occurs when iron acetate is added but the lake dies anyhow. Yet the maximum regret of adding the iron acetate is 5, while the maximum regret of doing nothing is 7. Yet unlike the motorcycle-helmet example, it is far less intuitively clear that precaution favors following minimax regret rather than maximin.

Table 3.4 *Hansson's example*

| | $s_1$ | $s_2$ | $s_3$ |
|---|---|---|---|
| Add iron acetate | 5 | 5 | 17 |
| Do nothing | 12 | 0 | 12 |

I believe that the explanation for this difference is that the harm of adding iron acetate in Hansson's example is not trivial in comparison to the harm it seeks to prevent. Indeed, it is just less than half of that harm. Consequently, it is questionable whether Table 3.4 could be reasonably approximated in qualitative terms in the manner of Table 3.3. In this case, plausible argument could be made that the harm resulting from both adding iron acetate and the death of the lake in $s_3$ deserves to be treated as qualitatively worse than the death of the lake alone. Moreover, this example illustrates that minimax regret is problematic from the perspective of MPP. For in a real case of the sort Hansson describes, estimates of the costs for various scenarios would most likely be uncertain, with a range of possible estimates being present. Therefore, when the harm imposed by the precaution is close to half of the harm it seeks to prevent, whether or not minimax regret recommends the precaution will be very sensitive to which cost estimate is chosen. For instance, if the cost of adding the iron acetate were assumed to be greater than 6 in Hansson's example, minimax regret would no longer recommend it. In such cases, then, minimax regret is susceptible to paralysis due to scientific uncertainty, in a manner similar to cost–benefit analysis (see section 2.3).

So what would PP recommend in a case of the sort described by Hansson? I suggest that it would challenge the artificial assumption that the only options are to add the iron acetate (in the specified amounts, according to the specified schedule) or to do nothing at all. For instance, why not begin by adding the iron acetate in small quantities and then make future decisions about whether to add more contingent on the observation of some positive response? In general, the search for more efficient alternative precautions is a central component of successful applications of PP. Moreover, as discussed in the next section, the search for such efficiency-enhancing alternatives is exactly what adaptive management is all about.

## 3.4 Robust adaptive planning

The concepts of robustness and adaptive planning – which I will for convenience jointly refer to as robust adaptive planning (RAP) – have been discussed in relation to PP by a number of authors (Doyen and Pereau 2009; Hauser and Possingham 2008; Johnson 2012; Mitchell 2009; Popper, Lempert, and Bankes 2005; Sprenger 2012). But there is much disagreement about whether RAP and PP are contrasting approaches or compatible and mutually supporting. In this section, I argue in favor of the latter option and explain how the concepts of consistency and efficiency, which lie at the core of proportionality, are closely linked to RAP and alternatives assessment as well.

### *3.4.1 Adaptive management and efficient precautions*

According to Bryan Norton, adaptive management's "first and defining characteristic" is experimentalism, which entails "taking action capable of reducing uncertainty in the future" (2005, pp. 92–3). A similar definition states that adaptive management consists of "learning through management, with management adjustments as understanding improves" (Williams 2011, p. 1371). Adaptive management, then, views policies as being implemented sequentially in such a way that earlier decisions should preserve flexibility of options for those that come later, and subsequent decisions should be informed by the consequences of earlier ones. Policies that possess these two features can be described as adaptive and can be contrasted with fixed strategies that enact an unchangeable plan.

From the perspective of the approaches considered in this section, adaptive management is not an end in itself. Instead, making a policy adaptive is a means of attaining some other goal, such as efficiency or robustness (see Lempert, Popper, and Bankes 2003, pp. 57–62). The robustness of a policy has to do with the extent to which its effectiveness is dependent on a specific future scenario being the actual one. The broader the range of plausible scenarios in which a policy achieves satisfactory results, the more robust it is. Robust policies are valuable from the perspective of PP because they can be justified by less demanding knowledge conditions, thus making the justification less susceptible to scientific uncertainty. Recall that MPP insists that the knowledge condition in an application of PP not be more stringent than necessary to satisfy consistency (see section 2.4.1). Consequently, strategies for finding more robust policies are extremely important for effective applications of PP.

I begin by discussing a proposal to interpret PP in terms of RAP advanced by Doyen and Pereau (2009). Doyen and Pereau propose to understand PP "in terms of a safety target that a decision-maker has to reach at a minimal cost in a robust way" (2009, p. 127). They understand robustness in a categorical manner: a policy is robust just in case it achieves a satisfactory, or "safe," result in every scenario under consideration. Doyen and Pereau develop a formal model that treats the decision as occurring in two temporal stages, wherein information is acquired at the second stage that reduces uncertainties. At each stage, the overriding aim of the decision maker is to ensure that the safety target is attained for that stage. The decision maker takes an action at each stage, which is represented in the model by setting a policy variable to a particular value. Doyen and Pereau distinguish two cases, which they term *strong precaution* and *adaptive precaution* (2009, p. 128). In strong precaution, the decision maker specifies in the first stage the values to which the policy variables will be set at that stage and the next. Hence, strong precaution does not involve learning at the second stage and corresponds to what was called a "fixed strategy" above. In adaptive precaution, the decision maker selects a value for the policy variable at the first stage but waits until new information is acquired before setting the policy variable at the second stage.

For the present purposes, there are two salient advantages of adaptive precaution in Doyen and Pereau's model. First, any situation in which strong precaution is capable of ensuring that the safety target is achieved in both stages is also one in which adaptive precaution can do so as well *but not the converse* (Doyen and Pereau 2009, p. 129). Since information is only gained and never lost at the second stage, if it is known from the start that a particular action ensures the safety target at the second stage then that is still known at the second stage. Hence, as far as achieving the safety target goes, nothing is lost by waiting until the second stage to make the decision. Moreover, since new information can be gained in the second stage, it is possible that a policy decision exists that ensures the safety target for every possible scenario remaining at the second stage but not for the full set present at the start. In short, adaptive management can result in more robust policies. Second, the cost of adaptive precaution is never greater, and sometimes less, than that of strong precaution (Doyen and Pereau 2009, pp. 129–30). This follows straightforwardly from the fact that strong precaution is a proper subset of adaptive precaution. Hence, adaptive precaution can generate more options for achieving the safety target and thereby the ability to achieve that target at lower cost.

Consider the relationship between Doyen and Pereau's proposal and the interpretation of PP advanced here. The harm condition in an application of PP corresponds to the complement of Doyen and Pereau's safety target. That is, an outcome satisfies the harm condition if it is not included among those that attain the safety target. Doyen and Pereau's first result, then, is an illustration of the precautionary virtue of robustness noted above. It shows how adaptive precautions can require less stringent knowledge assumptions for their justification, making them less vulnerable to challenges arising from scientific uncertainty. Meanwhile, Doyen and Pereau's second result illustrates that adaptive precaution is a means for promoting the efficiency of precautions.

Nevertheless, there are some differences between Doyen and Pereau's proposal and PP as interpreted here. Doyen and Pereau define PP as the set of those cases in which an adaptive precaution can ensure the safety target but no strong precaution, or fixed strategy, can (Doyen and Pereau 2009, p. 129). The interpretation proposed here does not restrict PP in this manner, and I see no plausible motivation for limiting PP in this way. On this point, I entirely agree with Thalos that there may be some circumstances in which a fixed strategy, such as putting on a motorcycle helmet, could be justified by PP. Another and more substantive difference is that Doyen and Pereau's approach cannot recommend anything if there is a scenario in which the safety target cannot be achieved by any action. In short, their approach is subject to the difficulty confronting the maximin rule discussed in section 3.3.3.

However, other RAP approaches exist that avoid this limitation, including one developed by a group of researchers at the RAND Corporation (Lempert, Popper, and Bankes 2003; Lempert *et al.* 2006; Popper, Lempert, and Bankes 2005). Unlike Doyen and Pereau, Lempert *et al.* treat robustness as a matter of degree. A policy is robust to the extent that it generates satisfactory results in a broad range of plausible scenarios. "Satisfactory" is defined via the concept of *regret* discussed in section 3.3.3: a policy achieves a satisfactory result in a given scenario if its outcome there falls within a specified acceptable range of the policy that achieves the best result in that scenario (Lempert, Popper, and Bankes 2003, pp. 55–6). Regret entails that dire scenarios – wherein catastrophe cannot be prevented by any action – will not paralyze the decision process. In fact, such dire scenarios have little, if any, influence on which policy is recommended, in agreement with the interpretation of PP proposed here.

This approach is also intended to be applicable even if no perfectly robust policy exits. It may be that, for every policy there may be a nontrivial vulnerability set of futures in which it performs very poorly. Two considerations are naturally prominent in such situations: the relative likelihood or scientific plausibility of the vulnerability sets and how poorly each policy performs in its vulnerability set (Lempert, Popper, and Bankes 2003, pp. 118–20). However, RAP does not attempt to provide an algorithm for deciding how to act in the face of such "irreducible risks" (Lempert,

Popper, and Bankes 2003, pp. 120–1). Instead, the emphasis is ultimately on characterizing the problem in a narrative form that makes the crucial uncertainties and trade-offs intelligible to decision makers. Considered in relation to the interpretation of PP proposed here, this version of RAP is, therefore, less prescriptive. In a case involving an unavoidable trade-off, PP as interpreted here directs one to find the least stringent knowledge condition that allows for some policy to be consistently recommended. Thus, PP aims to provide specific guidance about what to do and not merely to frame trade-offs in a perspicuous way.

In sum, RAP is relevant to PP insofar as it demonstrates how an adaptive approach to policy making can (i) facilitate finding a policy that can be consistently recommended by some version of PP even given substantial scientific uncertainty, and (ii) improve the efficiency of precautions. These two factors are also central to the relationship between PP and alternatives assessment, according to which effective policies for protecting human and environmental health should actively search for and incentivize feasible safer alternatives to materials, procedures, or activities currently in use (O'Brien 2000; Raffensperger and Tickner 1999; Tickner and Geiser 2004). Alternatives assessment can be helpfully understood by reference to a widely discussed distinction between active and passive adaptive management (see Holling 1978; Walters 1986; Williams 2011). An *actively adaptive* policy aims to accelerate the acquisition of knowledge that will point policy in one direction or another, while a *passively adaptive* policy simply waits for those signposts to appear. Thus, from the perspective of PP, alternatives assessment is driven by the insight that actively adaptive approaches to policy can further advance (i) and (ii).

### *3.4.2 Interpretation or alternative?*

As noted above, authors who discuss RAP in connection with PP frequently disagree about the relationship between the two. One perspective, defended here, is that they are compatible and that RAP helpfully clarifies some important aspects of PP. We have already encountered a view of this sort in Doyen and Pereau's proposal to explicate PP in terms of RAP. Alan Johnson (2012) also supports this position with a philosophical reflection on the link between precaution and intelligent tinkering à la Aldo Leopold (1966). A second viewpoint presents RAP and PP as conflicting approaches to decision making. For example, some equate PP with the maximin rule and point out that maximin is distinct from RAP (Popper, Lempert, and Bankes 2005), while others presume that PP entails a strict policy of non-interference that is contrary to adaptive management (Hauser and Possingham 2008). A third group takes an ambiguous stance. For instance, Sandra Mitchell argues in favor of RAP and suggests that it can "provide one interpretation of how to take a precautionary approach" (2009, p. 100). But she follows this statement with a critique of PP on the grounds that it unreasonably demands unequivocal proof of absence of harm, without explaining how a precautionary approach differs from PP (Mitchell 2009, pp. 100–1). Jan Sprenger (2012) offers a similarly ambiguous position on RAP and PP. Sprenger rejects PP as a decision rule, citing the theorem due to Peterson (2006) discussed in section 2.5.2 (Sprenger 2012, pp. 882–3). But he subsequently goes on to propose an interpretation of PP in terms of robustness that, to all appearances, is a decision rule (Sprenger 2012, p. 886). In this section, I respond to the arguments just listed about the incompatibility of RAP and PP and then explain why the ambiguous stance on RAP and PP is not coherent.

None of the reasons mentioned above in support of the claim that RAP and PP conflict are sound. Although PP coincides with maximin under the Rawlsian conditions, the two are not equivalent in general as explained in section 3.3.3. Nor is there any reason that PP must always recommend a policy of non-interference. For what if non-interference leads to catastrophe, while interfering can prevent catastrophe at a minimal cost? Nor does PP entail a blanket demand for unequivocal proof of absence of harm before allowing any activity to proceed. Indeed, such a demand would not be allowed by MPP, as illustrated in section 3.3.1. Finally, section 2.5.2 explains why Peterson's (2006) "impossibility theorem" does not succeed in establishing its charge against PP.

With this ground cleared, let us return to the ambiguous stance on RAP and PP. This approach appears to be motivated by the idea that PP can function effectively as a meta-rule, which places some general constraints on what sorts of decision-making procedures should be used in environmental policy, but not as a decision rule, which selects among policy options. This position is explicitly asserted in Sprenger (2012, p. 884), who cites Katie Steele (2006) in support of it. An examination of Mitchell's text suggests a similar interpretation. She argues that RAP is preferable to cost–benefit analysis on policy issues concerning complex systems, such as climate change, because RAP is not paralyzed by scientific uncertainty (2009, pp. 85–97). And her statement that RAP is one interpretation of a "precautionary approach" is preceded by a quotation of the famous Rio Earth Summit statement of PP, according to which scientific uncertainty should not be a reason for postponing cost-effective precautions

(Mitchell 2009, p. 100). Thus, the ambiguous stance is apparently motivated by the conviction that MPP expresses an important insight concerning environmental policy combined with skepticism of the rationality of PP as a decision rule.

However, this "meta-rule but not decision rule" position on PP is inherently unstable and ultimately incoherent. For if MPP is coherent and not trivial, then there must be coherent and informative decision rules that are compatible with its directives. But then decision rule versions of PP can be found among these decision rules that accord with MPP. For example, Sprenger proposes that PP might be understood in terms of robustness as follows: "An act that allows for certain environmental hazards has duly taken into account precautionary considerations if diverse, structurally different models indicate that these hazards may be negligible" (Sprenger 2012, p. 886). This statement is a proposed rule specifying when PP allows an action with possibly harmful environmental consequences to proceed. As such it is plainly a decision rule: the rule could be cited in support of a decision to permit that activity if its conditions were satisfied. The point here is not to advocate Sprenger's proposal but only to make the point that, if MPP has any value, some coherent and non-trivial decision rules that satisfy its strictures must be forthcoming. So, to defend MPP while denying the possibility of coherent decision rule versions of PP is to adopt an untenable position.

## 3.5 Conclusions

To have any clear sense of what PP asserts, it is necessary to impose some order on the rowdy zoo of interpretations of it that have been proposed. In this chapter, I have shown how the interpretation of PP advanced here does just this with regard to catastrophe principles, the maximin rule, minimax regret, robust adaptive planning, and alternatives assessment. In the next chapter, I turn to a positive argument for PP.

CHAPTER 4

# *The historical argument for precaution*

> Over the course of our lives, we develop strategies for avoiding the kinds of mistakes to which we are most susceptible.
>
> Philip Kitcher, *Science in a Democratic Society*, p. 113

## 4.1 Learning from history

In this chapter, I elaborate an argument for PP that rests on the history of environmental policy. This history contains many cases of prolonged and ultimately costly delays in response to serious environmental problems, while rushes to unnecessary and seriously harmful environmental regulation are relatively rare. The historical argument is not new. Indeed, it is hinted at in the Wingspread Statement on the Precautionary Principle (Raffensperger and Tickner 1999, p. 353) and is the focus of an edited volume published by the European Environment Agency (EEA) under the title, *The Precautionary Principle in the 20th Century: Late Lessons from Early Warnings* (Harremoës *et al.* 2002).[1] The cases discussed in this EEA volume are summarized in section 4.2.1. However, the historical argument for PP has yet to receive a thoroughgoing philosophical defense and, as a result, remains crucially incomplete in several respects. First, critics of PP can respond by listing countervailing cases of what they view as excessive precaution. The authors of the EEA volume state that they considered such cases but found them insufficiently "robust" to merit inclusion (Harremoës *et al.* 2002, pp. 3–4). But a defense of the historical argument requires explicitly examining a list of proposed cases of excessive precaution, which I do in section 4.2.2. Second, and perhaps of greater significance, is that a number of as yet unanswered objections have been raised to the logic of the historical argument for PP. Addressing these objections is the task of sections 4.3 and 4.4.

[1] Kriebel *et al.* (2001, pp. 871–2) also present a version of the historical argument.

In section 4.3, I respond to the objection that, even if the premises of the historical argument are granted, PP does not follow. Instead, the objection continues, the only rational response is an unwavering insistence that all types of risk be considered in an evenhanded fashion (Graham 2001; Hourdequin 2007), as risk trade-off analysis endeavors to do (Graham and Wiener 1995). In section 4.3.1, I respond that PP is compatible with an equitable approach in which risks of all types are accorded the weight they deserve. That is, the historical argument for PP is that environmental policy has *not* been evenhanded in its treatment of types of risk but instead is biased in the direction of prolonged delay in the face of environmental hazards. Thus, PP functions as a corrective intended to move policy making on environmental matters toward greater balance. In section 4.3.2, I make the case that, unlike PP, risk trade-off analysis offers no corrective for the historical imbalance documented in section 4.2.

Finally in section 4.4, I address an epistemic challenge to the historical argument for PP due to Munthe (2011, chapter 4). Munthe frames this objection as a critique of a two-level consequentialist argument for PP. Two-level consequentialism recommends that one follow rules that have been found to produce better results than alternative rules. Moreover, the historical argument for PP could be construed as an instance of this pattern. Munthe's objection, then, is that the epistemic challenges of predicting the consequences of the acceptance of a general rule are insurmountable, and therefore that a justification of PP must rest on some other basis. In section 4.4.1, I respond that the historical argument for PP need not be understood in terms of two-level consequentialism. I suggest that the argument rests on a simpler principle that imposes much less burdensome epistemic requirements, namely that if a systematic pattern of errors has occurred, then a corrective should be sought. In section 4.4.2, I critically examine the contention that the normative basis of PP must be sought in the controversial details of one or another ethical theory.

## 4.2 The case for precaution

In this section, I make the case that failures of governments and industry to adequately heed early warnings about harmful health or environmental effects have been more frequent and damaging than the opposite type of mistake, in which irrational fears lead to a rush to regulate a substance or activity that is later found to be harmless. This argument supports MPP, which emphasizes the importance of taking action in the face of environmental hazards despite uncertainties.

### 4.2.1 *Early warnings, late lessons*

There are numerous examples of environmental problems in which effective action was preceded by extended delays, during which time harms accumulated that ultimately resulted in a much greater mess to clean up than if prompt action had been taken. At present, climate change is the most obvious and disturbing case of this sort. But the history of environmental policy is replete with many additional cases: asbestos, polychlorinated biphenyls (PCBs), polybrominated diphenyl ethers (PBDEs), and leaded gasoline to name just a few. In most cases, the time lapse between the initial evidence of harm and effective regulatory action was a decade or more, and in some instances, as in the case of asbestos, it was over a century. In other cases, such as climate change, the clock is still ticking.

The fourteen cases of "late lessons from early warnings" assembled in the European Environment Agency volume cited above (Harremoës *et al.* 2002) are listed below, with a brief description for each.

1. *The collapse of the California sardine fishery*: State scientists in California had warned from the 1920s that rates of fishing were not sustainable, but these warnings never led to action and the sardine fishery collapsed in 1942, not showing signs of recovery until the 1980s (McGarvin 2002, p. 24).
2. *Radiation*: Reports of harmful effects of radiation begin just before the turn of the twentieth century, but no regulations concerning the use of radioactive materials (e.g., in workplaces) were enacted in the UK until 1961 (Lambert 2002, p. 33).
3. *Benzene*: Reports of benzene causing aplastic anemia date to 1897, while the first studies linking benzene to leukemia date to 1928. In the US, effective regulations on benzene were not introduced until 1977, and implementation of these regulations were delayed a decade further due to litigation (Infante 2002, p. 48).
4. *Asbestos*: Reports of harmful effects of asbestos on lungs date to 1898. Asbestos was banned in the European Union in 1999 (Gee and Greenberg 2002, p. 62). In the US, the EPA (Environmental Protection Agency) proposed a phase-out of asbestos in 1989, which was overturned in court.
5. *Polychlorinated biphenyls (PCBs)*: Reports of toxic effects of PCBs date to 1899, and reports of liver damage due to PCBs date to 1936. In the US, the Toxic Substances Control Act of 1976 restricted use of PCBs. In 1996, the EU decided to phase out PCBs by 2010 (Koppe and Keys 2002, p. 78).

6. *Halocarbons and the ozone layer*: In 1973–4, research on the distribution of chlorofluorocarbons in the upper atmosphere and a theoretical explanation of how they could destroy the ozone layer was published. An ozone hole was detected over Antarctica in 1985, leading to the Montreal Protocol on Substances that Deplete the Ozone Layer in 1987 (Farman 2002, p. 89).
7. *Diethylstilboestrol (DES)*: Reports of cancer in animals exposed to DES date to 1938. In 1953, a randomized controlled trial found DES to be ineffective for preventing miscarriage. In 1970–1, studies were published of vaginal cancer in women exposed to DES in utero. The US FDA (Food and Drug Administration) withdrew approval of DES in 1971 (Ibarreta and Swan 2002, p. 99).
8. *Antimicrobials as growth promoters in animal feed*: Awareness of the potential for evolution of resistance to antimicrobials by bacteria dates to 1945. The EU banned several antimicrobials in animal feed in the 1990s, and issued a general prohibition in 2006 (Edqvist and Pedersen 2002, p. 109). Antimicrobials are still commonly used as growth promoters in animal feed in the US.
9. *Sulfur dioxide*: Studies linking acid rain to sulfur dioxide emissions date to the early 1970s. A EU-wide sulfur protocol was negotiated in 1994 (Semb 2002, p. 120). In 1995, the US EPA implemented the Acid Rain Program, which established a cap-and-trade scheme to regulate sulfur dioxide emissions.
10. *Methyl tert-butyl ether (MTBE)*: Studies of the low biodegradability of MTBE in water date to 1954. Use of MTBE as a gasoline additive increased in the 1970s. In 1990, studies finding MTBE in groundwater appear (von Krauss and Harremoës 2002, p. 137). Twenty-five states in the US issued bans on MTBE in the 2000s.
11. *Organochlorine pesticides and the Great Lakes*: The publication of Rachel Carson's *Silent Spring* in 1962 drew public attention to the harmful effects of pesticides on wildlife. Dichlorodiphenyltrichloroethane (DDT) was banned in Canada in 1969, and in the US in 1972 (Gilbertson 2002, p. 146).
12. *Tributyltin (TBT)*: Increased use of TBT as an antifouling agent on the hulls of ships in the 1970s was associated with declines of marine snail and oyster fisheries. Individual states began instituting bans on TBT in the 1980s (Santillo, Johnston, and Langston 2002, p. 159). The International Convention on the Control of Harmful Anti-Fouling Systems on Ships was negotiated in 2001 and came into force in 2008.

13. *Growth hormones*: Concerns about estrogenic steroids used as growth promoters in livestock arose in the early 1970s. In 1982, a report commissioned by the EU concluded that some growth hormones posed no threat to human health, yet the EU issued a ban on growth hormones in 1985, which was implemented in 1988 (Bridges and Bridges 2002, p. 169). In some cases, refusal to approve growth hormones for agricultural use was made on the basis of concerns about adverse effects on animal health (see section 9.3).
14. *Bovine spongiform encephalopathy (BSE)*: The US banned scrapie-infected sheep and goats from animal and human food stocks in the 1970s. In 1989, the UK introduced an ineffective ban on cattle brains and offal in human food supply. In 1996, cases of Creutzfeldt–Jakob disease occurred in the UK (van Zwanenberg and Millstone 2002, p. 184).

In five of these cases (radiation, benzene, asbestos, PCBs, and antimicrobials as growth promoters in the EU), the delay between early warnings and effective action was more than fifty years. In three other cases (DES, MTBE, TBT), the lag was in the region of thirty years. About twenty years elapsed in the case of sulfur dioxide in the EU, while the delays in addressing organochlorine pesticides and depletion of the ozone layer were comparatively brief, lasting only about a decade in each case. In some instances, the episodes were brought to a close not by prudent regulation as much as by a collapse, as in the case of the California sardine fishery, or by a fiasco, as in the case of the UK and BSE. Finally, several of these cases remain pending, including antimicrobials as growth promoters in US agriculture and MTBE in the EU. The case of growth hormones, specifically recombinant bovine growth hormone, is discussed in further detail in the final chapter.

In addition to the time lags, the fourteen cases listed above illustrate another important concept often associated with PP, namely irreversibility. In almost all of these cases, the harm to human health or the environment persists for decades or more even after the activity generating the harm ceases. For instance, additional cases of mesothelomia continue to arise as a result of past and continuing asbestos exposure, and medical costs resulting from treating new and existing cases of illness caused by asbestos continue to accumulate, along with costs of clean-up or abandonment of asbestos-contaminated sites. In an extreme case, contamination from asbestos mining has transformed Wittenoom, Australia, into a ghost town unfit for human habitation for the foreseeable future. The concept of robust adaptive planning (RAP) is helpful for explaining the importance of irreversibility to PP. As discussed in Chapter 3, robust policies (i.e., policies that perform reasonably well in a wide array of plausible future scenarios) are often adaptive. An adaptive policy takes action now while avoiding as much as possible foreclosing future options that may be preferable in light of information acquired later. Thus, a policy that leads toward a potentially serious and irreversible harm is the opposite of adaptive, since such a policy commits future decision makers to choosing the least bad of a bad lot in the event that the potential harm materializes. From the perspective of PP, then, irreversibility is a red flag that signals the necessity of careful deliberation before proceeding.

### *4.2.2 Counterbalancing cases of excessive precaution?*

The fourteen cases listed above, then, illustrate how lengthy delays in response to emerging threats to human health and the environment have frequently resulted in serious and persistent harms and as such are a central motivation for PP. But an argument of this kind is incomplete without some consideration of the opposite possibility of costly and unnecessary environmental regulations. For example, Marchant and Mossman suggest that cases of insufficient precaution can be equally balanced with those of excessive precaution:

> Of course, other examples could also be cited where, in retrospect, perhaps too much precaution was applied to what turned out to be insignificant or nonexistent risks. Examples of excessive precaution include saccharin, silicone breast implants, electromagnetic fields, Bendectin, "ice minus" bacteria, the MMR (measles, mumps, and rubella) vaccine, and the effect of genetically modified Bt corn on the monarch butterfly. (Marchant and Mossman 2004, p. 8)

However, an examination of the cases listed by Marchant and Mossman fails to support their suggestion that rushes to unnecessary precautions are just as serious a problem as costly failures to take precautions soon enough.

In assessing these examples, I will focus on the EU and the US. I also focus on regulation by government agencies charged with protecting human health or the environment, because that is the type of action PP is intended to justify.

1. *Saccharin*: Health concerns about saccharin were based on studies finding a carcinogenic effect in rodents, which were subsequently judged not to extrapolate to humans due to physiological differences. Saccharin was never banned in the EU or the US, although the US

did require warning labels on foods containing saccharin from 1958 to 2000, and the US FDA did attempt unsuccessfully to ban saccharin in the 1970s.

2. *Silicone breast implants*: The US FDA implemented a voluntary moratorium on silicone breast implants in 1992 to allow time to investigate issues about safety and health effects.[2] During this time, silicone implants were allowed only for cases involving breast reconstruction or replacement of ruptured implants, with the requirement that the subjects be tracked as part of a data set maintained by implant manufacturers. The moratorium was lifted in 2006 due to improved designs, which decreased the likelihood of rupture and migration of silicone within the body and due to mounting research finding no link between silicone breast implants and a number of alleged adverse health effects, including fibromyalgia and rheumatoid arthritis (see Lipworth, Tarone, and McLaughlin 2004). However, manufacturers of silicone breast implants are still required to collect and analyze follow-up data of implant patients.
3. *Electromagnetic fields (EMFs)*: There are currently no federal regulations in the US that place limits on exposure to EMFs from power lines.[3] However, the US Federal Communications Committee (FCC) places an upper limit on the quantity of radiofrequency (RF) energy that may be emitted by handheld communication devices, such as cell phones.[4] Similar safety standards exist in the EU.[5] A review carried out under the auspices of the International Agency for Research on Cancer found that there is "limited evidence" for a carcinogenic effect of RF-EMFs in humans and recommended classifying RF-EMFs as "possibly carcinogenic to humans" (Baan *et al.* 2011, p. 625).
4. *Bendectin*: Merrell Dow Pharmaceuticals introduced Bendectin in 1956, a mixture of vitamin B6 and a commonly used antihistamine, doxylamine. In the 1970s, Bendectin was the target of a series of lawsuits filed by plaintiffs claiming that Bendectin had caused birth defects in their children. Although Merrell Dow ultimately prevailed

[2] See US Food and Drug Administration regulatory history of silicone breast implants in the US: http://www.fda.gov/MedicalDevices/ProductsandMedicalProcedures/ImplantsandProsthetics/BreastImplants/ucm064461.htm, accessed September 25, 2012.

[3] See the US EPA's webpage on electric and magnetic fields from power lines: http://www.epa.gov/radtown/power-lines.html, accessed September 25, 2012.

[4] See the US FCC's webpage on wireless devices and health concerns: http://www.fcc.gov/guides/wireless-devices-and-health-concerns, accessed October 16, 2012.

[5] See the European Commission's website on public health and electromagnetic fields: http://ec.europa.eu/health/electromagnetic_fields/policy/index_en.htm, accessed September 25, 2012.

against these legal challenges and never made any payments to the plaintiffs, it withdrew Bendectin from the market in the US in 1983 to avoid further legal expenses. However, Bendectin continues to be available in the EU, as well as Canada, under several alternative brand names.

5. *"Ice minus" bacteria* (Pseudomonas syringae *742RS)*: *P. syringae* is a species of bacteria that is involved in the formation of frost on plants and is also thought, possibly, to play a role in generating precipitation. So-called "ice minus" bacteria are a genetically engineered strain of these bacteria lacking surface proteins that enable the usual *P. syringae* to form ice crystals. The US EPA currently classifies *P. syringae* 742RS as a microbial pesticide,[6] which the EPA defines as "Microorganisms that kill, inhibit, or out compete pests, including insects or other microorganisms."[7]
6. *Measles, mumps, and rubella (MMR) vaccine*: Marchant and Mossman appear to be referring to the case of a research article published in the *Lancet*, a leading British medical journal, in 1998 that suggested a link between the MMR vaccine and autism. That article was later retracted due to evidence of fraud and conflicts of interest on the part of the lead author, Andrew Wakefield, who was subsequently censured and stripped of his license to practice medicine in the UK. However, Wakefield's article and public statements about the dangers of the MMR vaccine were followed by a temporary decline in vaccination rates in the UK and an associated increase of measles cases.
7. *Bt* (Bacillus thuringiensis) *corn and the monarch butterfly*: Bt corn is a genetically modified organism that incorporates genetic material from the bacterium *Bacillus thuringiensis* that produces a protein that is toxic to Lepidoptera larvae, but which is thought to be harmless to other insects and other species, including humans. Bt corn is useful for combating the European corn borer, which is difficult to kill with conventional externally applied pesticides because it burrows into the stalk of the plant. In the US, Bt corn was first registered with the EPA by several biotech firms in the mid-1990s, and it is now widely grown throughout the US and a number of other nations around the globe. Controversy arose in 1999 following published research showing that corn pollen could be toxic to monarch butterfly

[6] See the US EPA's webpage on registered biopesticides: http://www.epa.gov/oppfead1/cb/ppdc/2002/regist-biopes.htm, accessed February 2014.

[7] See the US EPA's webpage on types of pesticides: http://www.epa.gov/pesticides/about/types.htm, accessed February 2014.

larvae (Losey, Rayor, and Carter 1999). However, subsequent research concluded that harmful effects to monarch butterflies from Bt corn pollen was unlikely from real-world exposure levels (Sears *et al.* 2001). The cultivation of Bt corn is far more restricted in the EU than in the US and is currently banned within a number of EU countries.

Let us consider, then, whether these examples support Marchant and Mossman's contention that examples of delayed or inadequate environmental precautions are equally balanced by cases of excessive environmental or health regulations.

In two of the cases on Marchant and Mossman's list, no regulation whatever was enacted (MMR, Bendectin). Wakefield's claims linking the MMR vaccine to autism did not lead to any new regulation of the vaccine, or even to serious lawsuits against its manufacturers, so this cannot be cited as an example of an excessive application of precaution. The primary response of public health agencies in the EU and US to the MMR/autism controversy has been to reassure the public about the vaccine's safety. Similarly, Bendectin was never subjected to more stringent regulation as a result of the birth defects controversy, and Merrell Dow was ultimately victorious in all of the Bendectin lawsuits. Indeed, the major legal legacy of Bendectin litigation is the *Daubert* ruling, which erected significant evidential hurdles for plaintiffs in tort cases (Cranor 2006).

In three others – saccharin, EMFs, and *P. syringae* 742RS – some regulations were enacted but these regulations did not prohibit the use of the entity. For example, saccharin could continue to be used as a food additive during the period that warning labels were required in the US. Similarly, regulations concerning EMFs obviously have not prevented mobile electronic communication devices from becoming a ubiquitous feature of life. Finally, consider the US EPA's classification of *P. syringae* 742RS as a microbial pesticide. This entails some restrictions on its use. For instance, a firm seeking to market *P. syringae* 742RS as a frost reducer for agricultural purposes would need to register its product with the US EPA, a process that would involve carrying out studies concerning potential health and ecological effects.[8] Some critics of the EPA have claimed that classification as a microbial pesticide has prevented *P. syringae* 742RS from being brought to the market in the US and thereby is indirectly responsible for costs to US agriculture of approximately $1 billion annually due to frost damage (Miller 2010). However, this argument is rather dubious, since if *P. syringae*

[8] See the US EPA's webpage on regulating biopesticides: http://www.epa.gov/pesticides/biopesticides/, accessed February 2014.

742RS did indeed have such vast economic potential, it would presumably be worth it for a corporation to pay the up-front costs of registration. Since over seventy other microbial pesticides have been registered with the EPA by a variety of firms (Braverman 2010, pp. 74–7), registration is clearly not an insurmountable obstacle. If no biotech firm has sought to register *P. syringae* 742RS, therefore, this may be due to doubts about its profitability.

Whether the three cases discussed in the foregoing paragraph can appropriately be described as cases of "excessive precaution" depends in part on how that phrase is understood. On a minimal interpretation, "excessive precaution" would refer to any precaution that turns out not to be necessary, such as fastening your seatbelt on a trip in which you are not involved in a crash. Given this interpretation, saccharin would appear to be a case of excessive precaution, since the warning labels alerted consumers to a threat that, to the best of our current knowledge, has turned out to be nonexistent. As for EMFs and *P. syringae* 742RS, scientific uncertainties make it unclear at present whether the relatively mild regulations that have been enacted go beyond what is necessary to protect human health or the environment.

But the term "excessive" has a stronger connotation that the precaution was *unreasonable* given the information available when it was enacted and not merely seen to have been unnecessary with the benefit of hindsight. In this sense of the term, a precaution, such as wearing a seatbelt, may not be excessive even if the risk it seeks to mitigate, such as a car accident, is never actualized. Given this stronger interpretation, it is questionable whether the case of saccharin constitutes an excessive precaution, since evidence that a food additive produces a carcinogenic effect in rodents provides reasonable grounds for restricting its use. Similarly, even if *P. syringae* 742RS turns out to be harmless, it seems entirely reasonable that safety testing be performed to confirm that fact prior to its widespread commercial use in agriculture. However, I stick with the minimal interpretation of "excessive precaution," since it can be applied in a less controversial and question-begging manner.

Let us, then, consider the last two cases on Marchant and Mossman's list: silicone breast implants and Bt corn. Both of these cases do involve some fairly restrictive regulation. The moratorium enacted by the US FDA substantially restricted, although did not entirely prohibit, the use of silicone breast implants for a period of fourteen years. However, it is less clear that this moratorium, viewed with the improved vision of hindsight, was unnecessary. As noted above, research carried out during the moratorium period has generally exonerated silicone breast implants of some adverse health impacts, such as rheumatoid arthritis and compromised immune function. Yet other serious health concerns relating to silicone breast implants – particularly, those having to do with rupturing – were genuine and have in fact been addressed by improved designs developed during the moratorium period. Nevertheless, for the sake of argument I will grant that the silicone breast implant case can be counted as an example of excessive precaution. Finally, the case of Bt corn is perhaps the best example on Marchant and Mossman's list. Bt corn is presently banned by several EU nations, including Germany and France, although it is cultivated in some other EU states, such as Spain (Gómez-Barbero, Berbel, and Rodríguez-Cerezo 2008). Moreover, the scientific consensus at present appears to be that, despite some laboratory evidence of harmful effects on non-target insect species (Lövei, Andow, and Arpaia 2009), adverse health or environmental impacts due to Bt corn in real-world settings are likely very small in comparison to compensating advantages from the reduced use of conventional pesticides (Clark, Phillips, and Coats 2005; Gatehouse *et al.* 2011; Naranjo 2009).[9]

Combing through Marchant and Mossman's seven cases, then, we end up with the following tally:

- three cases that could plausibly be characterized as involving more precautionary regulation than necessary (saccharin, silicone breast implants, and Bt corn);
- two cases involving relatively mild precautionary regulations that may or may not turn out to be unnecessary (EMFs, *P. syringae* 742RS); and
- two cases involving no precautionary regulation at all (Bendectin, MMR).

Obviously, the last two cases provide no support for Marchant and Mossman's argument. Let us consider, then, how the severity of the harms in the other five cases compare to the fourteen examples of late lessons from early warning canvassed above. The harms in those fourteen cases included such things as the collapse of entire fisheries, 400,000 thousand deaths due to asbestos in the EU alone (Gee and Greenberg 2002, p. 58), and sources of groundwater rendered unusable by MTBE (von Krauss and Harremoes 2002, p. 124). In many instances, the harmful effects were not only severe but will persist for decades or more. And this did not include climate change, which threatens to top the list of disastrous cases of delayed responses to scientific warnings of environmental harm. In comparison, the harms in the saccharin and silicone breast implant cases

[9] Whether similar claims hold true of other genetically modified crops is not a question I explore here (cf. Pleasants and Oberhauser 2013).

appear absurdly trivial: drinking diet soft drinks sweetened with aspartame instead of saccharin and cosmetic breast enlargement using an implant filled with saline solution instead of silicone. Can one seriously claim that these effects are comparable to the adverse impacts of collapsed fisheries, benzene, asbestos, PCBs, MTBE, or climate change? Perhaps Marchant and Mossman would point to economic losses suffered by manufactures of silicone breast implants, which, for instance, led Dow Corning to file for bankruptcy in 1995 (Feder 1995). However, costs to manufacturers were primarily due to lawsuits, not to the moratorium (i.e., to the precautionary regulation), and the vast majority of these lawsuits were ultimately decided in the manufacturers' favor (Schleiter 2010). The costs in the cases of EMFs and *P. syringae* 742RS, even if they were genuine examples of excessive precaution, also appear quite modest: limits on the quantity of RF-EMFs that cell phones may emit and requiring a firm wishing to market *P. syringae* 742RS to perform initial safety testing of its product. Again, there is no plausibility whatever to the suggestion that these examples stand as an equal counterweight to asbestos or climate change.

Of the entries on Marchant and Mossman's list, the ban on Bt corn, which has been implemented in several EU member states, is the most serious example of an unnecessary precautionary regulation. This case involves economic losses to farmers who wish to adopt Bt corn, losses to the biotech firms that would sell it to them, and possibly higher food costs for consumers, as well as costs stemming from the greater use of conventional pesticides. For example, data from Spain, where Bt corn is cultivated, documents increased income and reduced pesticide use among cultivators of Bt corn (Gómez-Barbero, Berbel, and Rodríguez-Cerezo 2008). The case of Bt corn does show that unnecessary regulation is not an imaginary or trivial problem and illustrates why PP should be applied in a proportional rather than absolutist manner. In the case of Bt corn, an absolute ban is difficult to consistently recommend. Given the lack of evidence of harm, the version of PP invoked would have a very minimal knowledge condition, such as this one that was discussed in section 2.4.1.

4. If it is possible that an activity will lead to serious harm, then that activity should be prohibited.

Yet this version of PP could not consistently justify a ban on Bt corn, since it is possible that such a ban might also lead to serious harm for reasons mentioned above.

Despite being a more serious example than the other entries on Marchant and Mossman's list, two important features of the Bt corn case limit the severity of the harms involved. First, the harms are mostly restricted to the

EU member states in which the bans are enacted. That is an important difference from several of the fourteen cases of late lessons from early warnings (e.g., sulfur dioxide, halocarbons and the ozone layer, PCBs, and TBT), wherein local emissions of problematic substances create adverse environmental or health impacts that cannot be contained within national boundaries. Of course, climate change is an additional and powerful illustration of this pattern. Second, bans can be rescinded through political or legal actions. That again contrasts with emissions of persistent organic pollutants, such as PCBs, that continue to do harm long after their production has ceased, and greenhouse gases that, once emitted into the atmosphere, may stay there for a century or more. While a political decision can promptly undo a government edict, it cannot cause accumulations of PCBs in the food chain or dangerous levels of atmospheric $CO_2$ to quickly disappear.

A critic might respond that the above analysis shows only that Marchant and Mossman chose poorly in assembling their list of cases of excessive precaution. The critic might continue by pointing to a number of well-known and influential studies of the high costs of environmental regulations, especially those that target toxic chemicals (e.g., Morrall 1986; Tengs *et al.* 1995). Such studies, the critic might assert, constitute the research upon which claims about the perils of excessive precaution are based. However, the literature claiming to demonstrate the excessive cost and inefficiency of many environmental and health regulations has itself been subject to sustained critique (Ackerman and Heinzerling 2004, chapter 3; Ackerman 2008a; Hansen, von Krauss, and Tickner 2008; Oreskes and Conway 2010; Tickner and Gouveia-Vigeant 2005). One of these critiques finds that the two influential studies about risks of regulation just cited, "are full of regulations that were never adopted and, in some cases, never even proposed" (Ackerman and Heizerling 2004, p. 47). The inclusion of cases involving no actual precautionary regulation is therefore not an isolated defect of Marchant and Mossman's list but is also a characteristic of some seminal risks-of-regulation research.

## 4.3 Risks and risks

One objection to the historical argument developed above is that it does not argue in favor of PP but instead for an evenhanded approach according to which all types of risks merit equal concern (Graham and Wiener 1995; Hourdequin 2007). Moreover, this objection suggests an argument against PP that is reminiscent of the incoherence horn of the dilemma discussed in Chapter 2, namely that selectively focusing on one type of risk is irrational because it can lead to increased risks overall. I respond to this objection in two parts. In section 4.3.1, I explain that the historical argument does not in fact presume that harms to the environment or human health should, as a matter of principle, be weighted more heavily than other sorts of harms, such as economic ones. Instead, the historical argument is based on the observation that the history of environmental policy exhibits a pattern of extended and ultimately costly delay in the face of serious environmental hazards. In this context, it advances PP as a corrective. In section 4.3.2, I respond to the objection that, if a corrective is desired, it should come in the form of risk trade-off analysis, which explicitly seeks to reduce overall risk, rather than PP.

### *4.3.1 The precautionary principle as a corrective*

Marion Hourdequin objects that, even if the premises of the historical argument are granted, PP does not necessarily follow:

> One might grant that risks to human health and the environment have been overlooked in the past, that traditional risk assessment is overly narrow and leads to underestimation of key risks, and that protective actions are sometimes needed under conditions of uncertainty. Yet, even conceding these points, one might reject the precautionary principle. In particular, one might argue that it is irrational to place extra weight on human-caused risks to health and the environment; instead, we should be concerned to minimize overall risk. (Hourdequin 2007, p. 343)

In an effort to answer this objection, Hourdequin considers – but ultimately rejects – a variety of conceivable justifications for counting anthropogenic environmental harms for more than other concerns. According to the interpretation of PP advanced here, however, the presupposition of this effort is false: PP does *not* propose a lexical ordering according to which harms to the environment or human health are prioritized over everything else. As discussed in section 3.2, such a position would be difficult to square with proportionality and is, moreover, hard to defend on its own terms given the extent to which harms to the environment, health, and the economy are interconnected. Instead, the historical argument advances PP as a corrective against an entrenched systemic bias.

All of the examples canvassed in section 4.2 involve a trade-off between short-term economic gain for an influential interest against a harm that was uncertain or distant in terms of space or time (or all three). Carl Cranor (1999) suggests that several pervasive informational and political asymmetries make prolonged delays in taking precautions likely in such circumstances. While the commercial benefits of a new material may be readily apparent, its harmful impacts might become manifest in surprising ways and only after a long latency period (Cranor 1999, pp. 77–8). And even after a harmful effect has been well documented, it may be difficult to unambiguously link it to a specific cause. Political asymmetries often work in the same direction. Those who have a financial stake in preventing regulation of a substance or activity are often organized into trade associations and well positioned to lobby legislatures or agency representatives to act in their favor. In contrast, those who suffer the harmful effects are often a disparate collection of individuals lacking coherent or effective organization (Cranor 1999, pp. 78–9).

Of course, no principle of decision making could be expected to eliminate deep-seated structural asymmetries such as these. However, a blanket requirement that environmental regulations be justified by cost–benefit analysis exacerbates the situation. More specifically, such a requirement makes it difficult to justify meaningful precautions in situations involving trade-offs between short-term economic gains and harms that are uncertain or temporally delayed. There are two reasons for this. First, decisions involving environmental policies are beset by substantial scientific uncertainties. As a consequence, cost–benefit analyses supporting opposite recommendations can often be generated by from distinct sets of assumptions. Second, applications of cost–benefit analysis typically rely on discounting-the-future assumptions, which naturally have the effect of magnifying short-term concerns in comparison to long-term ones. Both of these points are relevant to the wildly diverging cost–benefit analyses of climate change mitigation discussed in section 2.3. A corrective, then, would propose a decision rule for environmental policy that avoids both of these shortcomings.

Chapter 5 explores the concept of uncertainty in detail, while the issue of future discounting is taken up in Chapter 6. In this chapter, I examine the role of MPP as a corrective for cases like those discussed in section 4.2.1. According to MPP, it is unwise to demand that a precaution be justified by procedures that are susceptible to making scientific uncertainty a reason for inaction in the face of serious threats to human health or the environment. And the survey of cases of "late lessons from early warnings" illustrates that the problem highlighted by MPP is real and significant. In this context, MPP functions as a corrective by advocating decision rules that are less prone to being paralyzed by scientific uncertainty. The emphasis on harms to human health and the environment in

MPP, then, should not be understood as an expression of a lexical ordering according to which environmental harms always trump economic harms. Moreover, such a lexical ordering is not entailed by the other two themes that, on my interpretation, constitute the core of PP. In particular, harm conditions in distinct versions of PP are ranked on the basis of their *severity*, not their *type*. Thus, in the discussion of climate change mitigation in Chapter 2, the focus of consistency is on the plausibility and severity of harms, not on whether those harms are classified as environmental or economic.

Some advocates of PP might feel that the corrective interpretation is not precautionary enough: isn't the whole purpose of PP to prioritize environmental over economic harms? But an emphasis on environmental/health concerns in comparison to narrowly economic interests can be interpreted as a call for a corrective rather than as an expression of a lexical ordering. That is, it can be understood as asserting that current practices tend to unreasonably downplay environmental and human health concerns while magnifying the short-term economic interests of certain influential groups, and hence that reform is required. Moreover, such an interpretation of PP is preferable, because a lexical ordering leads to absolutism. If environmental concerns always count for more than economic ones, then no economic cost is too high for the prevention of any environmental harm no matter how trivial. Such absolutism is implausible on its face as well as incompatible with proportionality, which is an established plank of PP as discussed in Chapter 2. Indeed, defenses of PP often dismiss absolutist interpretations as unfair distortions (see Godard 2000; Percival 2006; Sandin *et al.* 2002).

An advocate of a stronger PP might reply that the demand to "place extra weight on human-caused risks to health and the environment" need not be construed as a strict lexical ordering. Rather, the intended image might be one of putting a few additional stones on the environment/human health side of the balance. But this suggestion faces several serious difficulties, two of which I discuss here. First, without some reasonably specific explanation of what it means to add "extra weight," the proposal is empty. The second difficulty is, in my view, even more serious. The "extra weight" proposal faces the following dilemma: either it asserts that human health and environmental concerns should be accorded the weight that they merit in decision making or it recommends that those concerns be accorded *more* weight than they merit. If the first, it does not differ from PP as I interpret it. For in that case, the claim would be that environmental and human health concerns are often given *less* weight than they actually deserve in environmental policy making so that a corrective is called for. But if the second interpretation is the correct one, then the proposal is incoherent. For what possible reason could there be for giving any consideration in a decision more weight than it merits? Indeed, together with the evident shortcomings of a lexical ordering, this dilemma suggests that the corrective interpretation is the *only* tenable way to understand PP.

Finally, consider an objection to the idea that PP can effectively act as a corrective against the tendency for uncertainty as a cause of delay in the face of serious threats to the environment or human health. A skeptic might ask whether the possibility of generating analyses for opposite conclusions exists for PP just as it does for cost–benefit analysis. Can't one just get any result out of PP by ranking the qualitative knowledge and harm conditions in a manner designed to derive a desired conclusion? Granted, no decision rule can be rendered immune from being distorted by individuals who have an interest in doing so. Nevertheless, there can be little doubt that established rules can frame discourse in a manner that strongly influences how decisions are actually made. Indeed, the intense controversy surrounding PP is evidence of this very fact. For if the accepted decision rules made no difference, why would anyone bother to object to PP or conventional cost–benefit analysis or any other decision approach? Moreover, there are important differences between PP that make it less susceptible to paralysis by uncertainty than cost–benefit analysis, a point which is nicely illustrated by the climate change example discussed in Chapter 2. From the perspective of PP, the question of whether to implement measures to substantially reduce GHG emissions, such as a carbon tax, turns on the question of whether there is a scientifically plausible mechanism whereby those measures could lead to catastrophe. If the answer to that question is *no*, as appears to be the case, then PP says that those measures should be put into action. In contrast, basing the decision on cost–benefit analysis leads to the very different and more difficult to answer question of whether the expected benefits of the proposed GHG emission reduction measures are greater than their expected costs. Answering this question requires making inherently problematic and disputable quantitative predictions about the adverse effects of climate change and their timing. That the question posed by PP is easier to answer than the one posed by cost–benefit analysis is illustrated by the fact that economists who argue against climate change mitigation, such as Tol, often grant that mitigation would not lead to catastrophe (Nordhaus 2012; Schelling 1997, p. 10; Tol 2010, p. 91). The ability of PP to act in a corrective role is further defended in the following section.

### *4.3.2 Risk trade-off and precaution*

Even granting that PP should be construed as a corrective to a historical pattern of failures to respond adequately to environmental hazards, one might object that a different corrective is preferable. Rather than PP, the objection might continue, and the rational approach would be one that seeks to reduce overall risk, whether those risks stem from industrial by-products or environmental regulations. John Graham and Jonathan Wiener (1995) advocate an approach of this sort, which they call risk trade-off analysis, and argue that it is superior to PP (Graham 2001; Graham and Wiener 2008a, 2008b). In this section, I argue, first, that risk trade-off analysis reinforces rather than counteracts the historical pattern described in section 4.2. Second, I explain how PP, as interpreted here, incorporates central concepts of risk trade-off analysis while nevertheless maintaining a corrective role.

The central theoretical contribution of risk trade-off analysis is a typology of four ways in which an action intended to eliminate a risk might create other risks (Graham and Wiener 1995, pp. 22–5). For convenience, I refer to an action aimed at eliminating or reducing a target risk as a *precaution*. Graham and Wiener's typology, then, is as follows. In *risk offset*, a precaution generates risks of the same type in the same population as the risk it aims to eliminate. For example, a regulation banning the artificial butter flavoring diacetyl, known to cause severe lung damage given persistent exposure, might result in risk offset if the replacement flavoring carries similar risks. In *risk transfer*, the precaution generates risks of the same type in a different population. A good example of risk transfer would be regulations requiring taller smokestacks on coal-fired power plants: the risks from the emissions are not removed but simply dispersed across a broader geographical area. In *risk substitution*, a precaution generates risks of a different type in the same population. The "smog shade" example discussed below is a possible example of this type of situation. Finally, in *risk transformation*, the precaution creates a distinct type of risk in a distinct population. For instance, banning Bt corn due to concerns about potential adverse health effects for consumers could entail a greater use of pesticides and thereby increased cancer risks among agricultural workers. In addition, Graham and Wiener discuss a number of factors that should be considered in weighing one risk against another, such as probability, size of the affected population, timing of risks, and ethical issues concerning the distribution of harms, although they do not provide any indication of how such considerations should jointly operate to lead to a decision (Graham and Wiener 1995, pp. 30–6). Finally, Graham and

Wiener propose that it is sometimes possible to find *risk-superior* precautions that eliminate a target risk without creating any other risks (1995, pp. 37–40).

However, rather than being an evenhanded approach to reducing overall risk as its supporters claim, risk trade-off analysis is clearly biased against environmental regulation. While a fair approach to unintended consequences of regulation would emphasize *both potentially positive and negative effects*, risk trade-off analysis focuses exclusively on the negative side of the balance (Revesz and Livermore 2008, pp. 55–65). But unintended *positive* consequences of environmental regulations are certainly possible. They can occur when harmful substances or activities have a diverse array of adverse impacts in addition to those specifically targeted by the regulation. In such cases, a regulation aimed at mitigating one harmful effect may also unexpectedly reduce others. For example, a growing number of empirical studies suggest that the phase-out of leaded gasoline in the 1970s in the US strongly contributed to the decline in violent crime rates that began there in the early 1990s (Mielke and Zahran 2012; Nevin 2000, 2007; Reyes 2007; Wright *et al.* 2008). But unintended positive consequences of regulation are plainly not what advocates of risk trade-off analysis want to talk about. As a result, risk trade-off analysis typically functions as a recipe for finding creative ways of arguing against regulations aimed at protecting human health or the environment. This point is nicely illustrated by an example concerning ground-level ozone, more commonly known as smog, and ultraviolet radiation.

In 1997, after a five-year review, the US EPA issued a stricter national ambient air quality standard for ground-level ozone, requiring that it be no more than 0.08 parts per million averaged over an 8-hour time period. American Trucking Associations successfully challenged the new standard in the US Court of Appeals in 1999. Part of the decision was a remand in which the court required the EPA to consider the potential benefits of ground-level ozone in providing protection from ultraviolet exposure. In 2003, the EPA issued its final response to this remand, which gave its reasons for rejecting the potential shading effects of ground-level ozone as an argument against the new ozone standard. Since the other parts of the 1999 *American Trucking Associations* v. *EPA* decision had already been overturned by the Supreme Court, this meant that the 1997 standard could be put into effect. Thus, while the legal challenge launched by American Trucking Associations did not ultimately prevail, it did succeed in delaying the implementation of the new ozone standard for nearly six years. Moreover, the argument made by the legal team of American Trucking Associations concerning the benefits of smog is a straightforward application of risk trade-off analysis, as the hypothesized beneficial effects of smog would be an example of risk substitution. Indeed, Graham and Wiener steadfastly insist that the concept of risk trade-off applies to this case (Graham and Wiener 2008b, p. 487). And some discuss the case as a landmark victory in the movement to force regulatory agencies to pay greater attention to risk trade-offs (Marchant 2001b). We will return to this case study shortly. For now, the relevant point is that the example illustrates the tendency of risk trade-off analysis to prolong regulatory delay in the face of hazards to human health or the environment.

In contrast, PP readily accommodates risk trade-off concepts while maintaining its role as a corrective against regulatory paralysis. Proportionality, construed in terms of consistency and efficiency, is motivated precisely by the possibility that trade-off effects may be associated with a precaution. Consistency requires that such effects do not rise to the levels of the knowledge and harm conditions in the version of PP used to justify the precaution. Efficiency requires that, among precautions that can be consistently recommended, those with less serious trade-off effects should be preferred. However, there is nothing in PP that biases the analysis toward a focus solely on unintended negative consequences of regulations. The potential of foregoing positive side effects of precaution could be a reason against maintaining the status quo. Indeed, the idea that adverse effects may propagate through complex systems in surprising ways is surely one motivation for PP, which therefore encourages research into the potential of such effects. Furthermore, unlike trade-off analysis, PP provides guidance about when potential trade-off effects should and should not be reasons against a precaution. Consider this point in the case of the "smog shade" example described above. Proportionality supports a straightforward argument for judging the potential shading effects of smog as not being a good reason against the ozone standard proposed by the EPA in 1997. First, while the harmful effects of ground-level ozone are both well established scientifically and substantial, evidence for protective shading effects of smog is weak and the effect is likely to be small even if it exists (Environmental Protection Agency 2003, pp. 627–36). Reasons for the likely small effect include the fact that ground-level ozone makes up less than 1% of the total ozone column and, unlike stratospheric ozone, is extremely variable by location and time of day. Thus, the potentially beneficial shading effect of ground-level protection from smog does not plausibly lead to a violation of consistency in this case.

In addition, efficiency generates an obvious argument for not taking the "smog shade" challenge seriously from the start. Naturally, efficiency favors ultraviolet protection measures that do not cause serious respiratory ailments to large numbers of people. Moreover, such measures (e.g., sunscreen, behavioral changes, protective clothing, sunglasses, measures to prevent depletion of stratospheric ozone) exist and are already widely advocated by government agencies concerned with public health and the environment. Thus, the sensible policy in this situation is to combine reductions of ground-level ozone with the promotion of benign protections from ultraviolet exposure. In comparison, it would be absurdly inefficient to advocate smog as a means of protecting the public from excessive exposure to ultraviolet light. I suspect that such reasoning explains the reluctance of the EPA to seriously consider the "smog shade" effect in their original justification of the 1997 ozone standard (cf. Marchant 2001b).[10]

## 4.4 An epistemic objection

A different objection to the historical argument focuses on epistemic obstacles inherent in drawing inferences about the likely effects of different rules, especially in an arena as vast and complex as environmental policy. If uncertainty undermines attempts to quantitatively predict the costs and benefits of a specific policy, the objection claims, then uncertainty must be even more debilitating when it comes to judgments about the effects of general decision-making methods. In this section, I consider a version of this objection advanced by Munthe (2011). In section 4.4.1, I reply that Munthe's objection unnecessarily links the historical argument for PP with a two-level consequentialist outlook. Instead, the historical argument rests on the premise that a systematic pattern of past errors calls for a corrective, which imposes far less stringent epistemic demands. In section 4.4.2, I consider the implications of this point for the claim that PP must be justified by appeal to deep principles of ethical theory.

### 4.4.1 *Precaution and two-level consequentialism*

First, a bit of background about the philosophical usage of the term "consequentialism" and a number of derivative concepts is in order. Consequentialism is the idea that rightness or wrongness should be judged solely in terms of consequences. Philosophers have teased apart many subtypes of

10 Hansen and Tickner make similar critical remarks regarding the "smog shade" case (2008, p. 476). However, the concepts of consistency and efficiency clarify the logic of the critique and are helpful for answering objections. For example, Graham and Wiener (2008b, p. 487) object that advocates of PP cannot dismiss the smog shade effect as merely "hypothetical" because PP is intended to justify precautions against hypothetical dangers. The analysis here shows the error of this objection. The relevant knowledge condition in an application of PP is not fixed across the board, but instead is set in relation to the available knowledge concerning the target risk.

consequentialism. I list below only those that are relevant for understanding the epistemic objection raised by Munthe to the historical argument for PP:

- *Consequentialism as criterion*: Consequentialism is the criterion, or standard, for what is right and wrong.
- *Consequentialism as a decision rule*: Actors should decide which action to take by tallying the positive and negative consequences of each possible action and selecting the one with the best balance of pros and cons.
- *Direct consequentialism*: The rightness or wrongness of an act is determined solely by the consequences of that act itself.
- *Actual (or factualistic) consequentialism*: Only *actual* consequences matter in assessing the rightness or wrongness of an action.
- *Expected consequentialism*: It is the likely or expected consequences that matter in judging whether an act was right or wrong.
- *Two-level consequentialism*: Accepts direct consequentialism as a criterion but advises against its use as a decision rule in most circumstances. Instead, two-level consequentialism recommends that decisions normally be guided by rules that, from experience, are known to generally produce better consequences than alternative rules.

The last two varieties of consequentialism can be viewed as a response to difficulties confronting the proposal that direct actual consequentialism should be used as a decision rule. The problem is that actors are almost never in a position to know all of the consequences of their actions, and hence efforts to apply direct actual consequentialism as a decision rule would result in either paralysis or self-serving rationalizations. Expected and two-level consequentialism both attempt to ameliorate this problem. Thus, according to expected consequentialism, a person might be in a position to know the likely consequences of her act even if the actual consequences cannot be known with certainty. In contrast, two-level consequentialism simply drops consequentialism as decision rule and proposes that decisions should be guided by rules that have, in past experience, tended to produce better results than others.

The historical argument recommends PP on the grounds of a history of insufficient precaution in the face of serious environmental and human health hazards. It is tempting, then, to view the historical argument as an example of two-level consequentialist reasoning, that is, as attempting to show that PP would lead to better results overall than other decision-making procedures that could be applied in environmental policy. Munthe's epistemic objection to the historical argument turns on just this interpretation:

> Hard as it may be to know what particular action is (factualistically) morally right when we lack the relevant information in a particular situation, to identify what manner of practical decision making would be right to apply as a general approach to risky decisions seems exponentially more difficult if this is to be done on the basis of a factualistic ethical theory. Applying any minimally plausible factualistic criterion of moral rightness and wrongness, we will never know what approach to risks will best meet this criterion (since risky situations are characterised by exactly that feature of lack of knowledge regarding what should be done according to a factualistic criterion of rightness). In addition, there is quite a lot at stake when choosing what decision making procedure to use, since the wrong choice may lead to catastrophes of enormous proportions, no matter what factualistic criterion of rightness we take as the basis for our choice. In effect, the choice of practical approach to risky decisions will itself be a risky decision of sorts, and one to which the requirement of precaution would seem to be clearly applicable. At least, we are in no position to assume the opposite. However, since – according to the suggestion at hand – we do not know what the requirement says before we have made our choice of approach, we cannot know what approach to choose. Thus, the two level approach cannot solve the problem of morally justifying any particular version of the requirement of precaution. (Munthe 2011, p. 63)

In short, Munthe argues that a two-level consequentialist argument for PP faces insurmountable epistemic challenges.

However, the historical argument for PP need not be understood as an application of two-level consequentialism. There is, in fact, a much simpler way to understand the logic of the historical argument that does not depend on a specific resolution of long-standing disputes in ethical theory. That alternative is to take the historical pattern discussed in section 4.2 as making the case for a need of a corrective so that the following premise is sufficient: *If a systematic pattern of serious errors of a specific type has occurred, then a corrective for that type of error should be sought.* In other words, the historical argument for PP rests on the commonsensical idea that rationality requires learning from mistakes. Although that idea is surely *compatible* with two-level consequentialism, it by no means *presumes* two-level consequentialism or any other controversial ethical theory.

Moreover, the epistemic demands on the historical argument for PP given the learn-from-mistakes interpretation are much less onerous than if the argument is construed as an application of two-level consequentialism. Given a two-level consequentialist approach, one would have to argue that the actual or expected difference of benefits over costs of adopting PP in environmental policy is greater than that of adopting any other rule. In contrast, given the learn-from-mistakes interpretation, the primary evidential burden is to demonstrate the existence of the systematic pattern of prolonged and ultimately costly delays in the face of serious yet uncertain threats to human health or the environment. Section 4.2 discussed supporting evidence for this pattern. Once the imperative for a corrective has been established, one can turn to a consideration of the merits of proposed correctives. For example, a proportional PP is preferable to an absolutist one for a number of reasons that have been discussed and, as explained in section 4.3.2, risk trade-off analysis provides no discernable corrective at all.

### 4.4.2 *Ethical theory and normative underpinnings*

The historical argument for PP, then, does not presume any deep or controversial ethical theory, which distinguishes it from a number of other defenses of PP that explicitly adopt a non-consequentialist perspective (Hansson 2003; John 2010; Kysar 2010; Munthe 2011). But can a substantive argument for PP really avoid such theoretical commitments? In this section, I critically examine arguments that it cannot.

Munthe advances an argument of this sort that rests on a distinction between instrumental rationality and moral obligation, as follows (Munthe 2011, p. 49). Instrumental rationality aims to find the best means to attain a specified end. But that a particular method is effective for an end (say, risk reduction) does not entail any obligation to abide by the directives of that method. For if we are not morally obliged to pursue that end and do not desire to do so, then the instrumental effectiveness of the method is irrelevant. Thus, the argument concludes, a positive case for PP must rely on one substantive ethical principle or another. In the case of the historical argument, Munthe might suggest, the most obvious suggestion is two-level consequentialism. Perhaps some other ethical theory could be used instead, but in any case it must be one capable of supporting strong moral claims.

However, the above argument is problematic because instrumental rationality and morality diverge primarily when one's aims are not moral – for instance, if my sole objective is to maximize my personal wealth. When the aims in question are ones that one is morally obliged to pursue, instrumental rationality and moral reasoning coincide much more closely. In the case of the historical argument, the errors in question involve harms to human health and lost environmental resources that are important to human livelihood. Surely, there can be no serious dispute that such things are of grave moral significance and that a powerful moral obligation exists to avoid errors that cause such effects. The historical argument needs nothing more than this by way of a moral basis. Or, to put the matter another way, it is true that an argument for PP requires substantive moral premises, but these premises need not take the form of controversial propositions of meta-ethics.

Another defense of the necessity of an ethical foundation of PP might point to the importance of equity issues in evaluating risks: "it makes a big difference if a person risks her own life or that of somebody else in order to earn a fortune for herself. Therefore, person-related aspects such as agency, intentionality, consent, equity, etc. will have to be taken seriously in any reasonably accurate general format for the assessment of risk" (Hansson 2009, p. 431; cf. Hansson 2003, p. 302). Although I agree with the conclusion of this argument, I think that it can be accommodated by insisting that environmental injustices play an important role in assessments of the severity of potential harms in applications of PP. But PP is not itself a theory of environmental justice. Nor should PP be saddled with the unnecessary baggage of claiming that consequentialist ethical theories are fundamentally incapable of providing an adequate theory of justice.

The argument for PP I develop here, then, has the attraction of avoiding reliance on debatable assumptions about ethical theory. Consider this in comparison with Munthe's proposal that PP is grounded on the conception of an irreducible responsibility to avoid introducing risks (Munthe 2011, chapter 5). This responsibility is *irreducible* in the sense that it is a moral obligation in its own right that is not justified on the basis of anything else, such as actual harmful consequences of risky behavior. The responsibility to avoid imposing risks is not absolute but comes in degrees (Munthe 2011, p. 91). Munthe also introduces a number of formal requirements relating to responsibility: for instance, that a decision is irresponsible if and only if there is some other decision that is more responsible (ibid.). In addition, he proposes the following necessary condition for responsible risk taking:

7. If a decision to introduce a risk is to be responsible, this decision must either produce some sufficiently substantial benefit, or sufficiently reduce some risk. (Munthe 2011, p. 92)[11]

Munthe then explains how this approach could be applied to several cases (2011, chapter 6).

Munthe's defense of PP assumes a non-consequentialist perspective on ethical theory, because it assumes that the responsibility to avoid imposing

[11] Munthe's proposition 6 appears to be logically equivalent to his proposition 7. See section 3 of the Appendix for an explanation.

risks is not justified by the appeal to consequences. Such a stance drops PP into the great philosophical debate on this topic that stretches back to Immanuel Kant, John Stuart Mill, and beyond. To see that the plunge into this morass is unnecessary, note that something very similar to Munthe's proposition 7 follows directly from the account of PP proposed here. If an activity creates a serious risk with no compensating benefit, then a version of PP will consistently recommend against it.[12] Thus, the defense of PP proposed here encompasses Munthe's proposition 7, while avoiding the controversial ethical-theory baggage.

## 4.5 Conclusions

It is not unusual for the history of belated responses to environmental hazards to be offered as a positive argument for PP. However, a thorough defense of this reasoning requires a careful examination of its logic and philosophical presuppositions so that several central objections can be addressed. Two important clarifications of the historical argument for PP have emerged from this process. First, rather than making an across-the-board claim that harms to health and the environment deserve more weight in decision making than other types of concern, the argument is best understood as aiming to show the need for a corrective. Second, the historical argument need not rely on controversial premises about ethical theory, such as the claim that consequentialism is preferable to deontology or vice versa. Instead, it rests on the premise that rationality involves learning from mistakes and hence that a history of significant errors generates an imperative to seek a solution.

[12] See Appendix 4 for a formal explication and proof of this claim.

CHAPTER 5

# Scientific uncertainty

## 5.1 Uncertainty about uncertainty

The concept of scientific uncertainty occupies a prominent position in nearly every statement of PP. But what, exactly, is scientific uncertainty? Uncertainty is often defined through a distinction with risk (Luce and Raiffa 1957, p. 13). Decisions under risk are ones in which possible outcomes of available actions are known and, although it is not known which outcome will actually occur, probabilities can be assigned to each. Decisions under uncertainty, by contrast, are those in which the possible outcomes of actions are known but not their probabilities. However, this definition is less transparent than it may seem on the surface because "probability" can be interpreted in more than one way. Probabilities might represent a stochastic process, such as a casino game or incidences of lung cancer, or they might represent the cognitive state of some actual or ideally rational person. I refer to the former sort of probabilities as objective chances, and the latter sort as personal probabilities. While risk is often defined in terms of objective chances, the distinction between objective chances and personal probabilities is often not clear-cut in practice (Elliott and Dickson 2011). And how probabilities are understood has major implications for the pervasiveness of risk versus uncertainty. On the one hand, if probabilities are interpreted strictly as objective chances, then it is doubtful whether any genuine cases of risk exist outside of casinos (Hansson 2009). On the other, if interpreted as personal probabilities, then almost any example in which possible outcomes are known can be treated as a case of risk. All of this leaves the significance of "scientific uncertainty," and hence the scope of PP, rather unclear.

In this chapter, I take a closer look at the concept of scientific uncertainty. In section 5.2.1, I argue that there are good reasons for not adopting the standard risk versus uncertainty distinction as a basis for interpreting PP. While this distinction is widely cited, it corresponds neither to the ordinary concept of uncertainty nor to the way in which uncertainty is understood in risk analysis. In risk analysis, risk and uncertainty are not treated as mutually exclusive decision situations, and probability is commonly used to quantify uncertainty. Moreover, I argue that, in ordinary usage, the outcome of some process is said to be uncertain when knowledge that would enable it to be predicted is lacking. In section 5.2.2, then, I propose to define *scientific* uncertainty in terms of the absence of a model whose predictive validity for the type of question at issue is empirically well confirmed. This proposal has a number of advantages over the decision-theoretic definition. First, it corresponds more closely to commonsense understandings of uncertainty, as well as to the concept of uncertainty found in risk analysis. It does not treat risk and uncertainty as mutually exclusive, it does not make knowledge of all possible outcomes of available actions a precondition for uncertainty, and it entails that scientific uncertainty may be present even when probabilities are known. These features are advantageous, because a definition of uncertainty that departs significantly from common usage is likely to lead to confusion and miscommunication. Second, since it does not define scientific uncertainty in terms of ignorance of probability, the definition proposed here avoids difficulties related to the interpretation of that concept. Finally, its rationale – that uncertainty about outcomes has to do with the inability to predict them – is a better fit for PP.

Since the extent to which the predictive validity of a model is empirically supported is a matter of degree, scientific uncertainty does not always entail that quantitative forecasts should be elided altogether in the decision-making process. This raises a number of important questions about the relationship between PP and quantitative approaches to environmental policy, such as risk analysis and cost–benefit analysis. In section 5.3.1, I suggest that PP and quantitative risk analysis can be used together in several ways. However, I also argue that PP is distinct from risk analysis in two respects. First, PP aims to provide a rule to guide decision making while risk analysis is primarily in the business of forming quantitative descriptions of risks. Second, PP is intended to be applicable to cases in which the extent of scientific uncertainty is sufficiently great to make quantitative risk assessments meaningless. Finally, in section 5.3.2, I argue that the compatibility of PP and risk analysis does not entail a similarly congenial relationship between PP and cost–benefit analysis. The sticking point I discuss is value commensurability, according to which costs and benefits should be measured on a common monetary scale. I argue that value commensurability is plausible neither as a psychological hypothesis nor as a pragmatic assumption aimed at making reasoning more transparent.

## 5.2 Defining scientific uncertainty

Section 5.2.1 critically examines the standard definition of uncertainty as knowledge of possible outcomes of actions but ignorance of their probabilities, while section 5.2.2 develops and defends an alternative proposal linking scientific uncertainty of outcomes to an inability to predict them.

### *5.2.1 The decision-theoretic definition*

According to a classic text on decision theory, *Games and Decisions*, uncertainty arises when one is faced with a decision between two actions wherein "either action or both has as its consequence a set of possible specific outcomes, but where the probabilities of these outcomes are completely unknown or are not even meaningful" (Luce and Raiffa 1957, p. 13). In decisions under risk, by contrast, both the set of possible outcomes and their probabilities are known. In what follows, I will refer to this as the *decision-theoretic definition* of risk and uncertainty. The decision-theoretic definition is often cited in literature on science and policy (see Elliott and Dickson 2011; Hansson 2009; Kysar 2010, p. 73; Shrader-Frechette 1991, pp. 101–2; Stirling and Gee 2003; Trouwborst 2006, pp. 86–9). It is often supplemented with a concept of ignorance, defined as both unknown probabilities and unknown possible outcomes. Other discussions add further wrinkles to the basic decision-theoretic structure, for instance, including disagreements about how to quantify utilities or how to frame the scope of the decision problem within the ambit of uncertainty (see Hansson 1996, p. 370; Lempert, Popper, and Bankes 2003, p. xii; Steele 2006, p. 27).

However, the decision-theoretic definition is unclear in at least two ways. First, it is not specified what sorts of probabilities are required to be known. Second, the final clause suggests that in some circumstances probabilities might not be "meaningful," but the definition does not explain the difference between "meaningful" and "non-meaningful" probabilities. Both of these concerns are connected to the issue of the interpretation of probability. Mathematically, probability is a function that satisfies a few formal criteria (Billingsley 1995; Stirzaker 2003): for instance, the probability of the entire outcome space must equal 1. But the definition of a probability function says nothing about what, if anything, probabilities represent. An interpretation of probability, then, proposes to answer this question. For the present purposes, interpretations of probabilities can be usefully divided into two main groups: *objective chance* interpretations, according to which probabilities represent some stochastic process operating in the world, and *personal probability* interpretations, according to which probabilities represent the degrees of confidence of some actual or ideally rational person. To illustrate the difference between objective chances and personal probabilities, consider a probability assigned to the proposition that the top seeded player will win her first round match in a tennis tournament. An objective chance in this case might be understood in terms of the relative frequency of cases in which top seeded players emerge victorious in first round matches in this tournament. In contrast, a personal probability would represent the degree of confidence of some person in the proposition that the top seeded player will win.

Given the decision-theoretic definition, whether an objective chance or personal probability interpretation is chosen has significant implications for the extensiveness of uncertainty versus risk. It is often easier to measure a person's degree of confidence than it is to empirically estimate physical propensities or relative frequencies. Indeed, given a personal probability interpretation, it would be difficult to understand how uncertainty could exist at all. According to the decision-theoretic definition, uncertainty requires that all possible outcomes of the available actions are known but not their probabilities. But given the list of possible outcomes, personal probabilities could be estimated simply by eliciting the degrees of confidence of some appropriate person, such as an expert. Thus, it would seem that an objective chance interpretation of probability must be what is intended in the decision-theoretic definition of uncertainty.

An earlier classic account of the difference between risk and uncertainty, due to Frank Knight, supports this interpretation (Knight 1921; cf. Elliott and Dickson 2011). Knight ties risk to a frequency interpretation of probability. The fundamental concept of such interpretations is the concept of relative frequency. The relative frequency of an event A with respect to a reference class R is simply the number of instances of A occurring in R divided by the total. The probability of an event, then, might be defined as its actual relative frequency in a large reference class or as the limit of its relative frequency in a hypothetical infinite sequence of cases. According to Knight, risk exists with regard to a class of events that, although not entirely homogeneous in all relevant respects, are sufficiently similar to be treated as a reference class for the purposes of analysis, thereby enabling an empirical estimation of the event's relative frequency. By contrast, uncertainty arises with respect to events wherein such a reference class does not exist, for instance, due to the uniqueness of the event. In such cases, probabilities come only in the form of personal degrees of confidence.

If knowledge of probabilities construed as relative frequencies is the benchmark of risk, then the "meaningful" probabilities alluded to in the definition of uncertainty cited above might be understood by reference to the concept of calibration (Hansson 1993, pp. 20–2). A probability assigned to an event is *calibrated* just in case it corresponds to the relative frequency of that event in a large sample of repetitions. For example, a meteorologist's statement that there is a 50% chance of rain is calibrated if it rains about half of the time when that forecast is made. Probabilities interpreted as relative frequencies are calibrated if and only if they true or at least approximately so. In contrast, personal probabilities could fail to be calibrated despite being true in the sense of accurately representing the cognitive state of the person in question. A probability claim would be meaningful, then, if there is some relative frequency to which it could be calibrated.

Such, then, is the conception of scientific uncertainty that one typically encounters in discussions of PP. However, it is doubtful that this concept of uncertainty is widely used in risk analysis and related endeavors, such as cost–benefit analysis. For instance, Terje Aven, a professor of risk analysis and risk management at the University of Stavanger, Norway, writes the following of his formative experiences as a practicing risk assessor who began reading up on the theoretical foundations of his discipline:

> References were made to some literature restricting the risk concept to situations where probabilities related to future outcomes are known, and uncertainty for the more common situations of unknown probabilities. *I don't think anyone uses this convention and I certainly hope not.* It violates the intuitive interpretation of risk, which is closely related to situations of unpredictability and uncertainty. (2003, p. x, italics added)

Thus, according to Aven, in both ordinary understandings of risk and the practice of risk analysis, risk and uncertainty are not treated as mutually exclusive decision situations, as the standard definition would have it. Instead, taking uncertainty into account is a normal part of assessing risks (Aven 2003, chapter 4; Bedford and Cooke 2001). Furthermore, rather than being defined as the absence of probability, uncertainty is often quantified by means of a probability distribution. In cost–benefit analysis, for instance, it is not unusual to first generate predictions of costs and benefits with a deterministic model. Uncertainty is then taken into account by introducing a probability distribution over some important inputs or parameters of the model and comparing the results of the probabilistic model to the original deterministic one. This practice is illustrated by some of the economic analyses of climate change mitigation discussed in section 2.3 (see Nordhaus 2008, pp. 123–5).

I think Aven is entirely correct that, in ordinary usage, uncertainty is understood as an aspect of risk rather than something that exists only when risk is absent. In addition, the decision-theoretic definition of uncertainty departs from the ordinary concept in several other significant ways. First, the decision-theoretic definition makes knowledge of all possible outcomes of potential actions a necessary condition of uncertainty. Yet it would surely be odd to assert, for instance, that the effects of climate change mitigation are not uncertain because there may be some outcomes that have not been anticipated. To the contrary, the potential for surprises is more naturally regarded as a factor that intensifies uncertainty. Second, knowledge of probability of outcomes is not normally regarded as sufficient for eliminating uncertainty. For example, one is uncertain which slot in a roulette wheel the ball will land in, despite knowing that the ball has a uniform probability of 1/38 of landing in each. The practice in risk analysis of treating probability as a means for quantifying uncertainty seems much closer to ordinary understanding in this regard. For example, consider a 95% confidence interval associated with an estimate of the expected number of fatal accidents per year if the speed limit on highways is increased by 5 mph. The typical assumption in such cases is that the wider the confidence interval, the greater the uncertainty. Thus, the estimate is more uncertain if the 95% confidence interval is ±1,000 rather than ±10. But in either case probabilities are known, and it does not appear to matter how those probabilities are interpreted, for instance, as frequencies or degrees of confidence. Indeed, the term "aleatory" or "stochastic uncertainty" would be used if the probabilities were objective chances and "epistemic uncertainty" if they were personal probabilities (Aven 2003, p. 17).

The definition of uncertainty as knowledge of possible outcomes but not their probabilities might be useful in decision theory despite diverging from ordinary usage. However, these striking differences call into question whether the concept of scientific uncertainty present in PP should be construed according to the decision-theoretic definition. Indeed, it is more plausible that drafters of various statements of PP would have had something closer to the ordinary concept of scientific uncertainty in mind. Moreover, using a concept of uncertainty that is so sharply at odds with ordinary understandings of the term is very likely to lead to confusion and miscommunication. But to judge whether PP is better interpreted in terms of a concept of scientific uncertainty that tracks ordinary usage more faithfully, it is first necessary to explicate its ordinary meaning more clearly.

In common usage, uncertainty implies an absence of knowledge. And like the decision-theoretic definition, PP is concerned with uncertainty in relation to the outcomes of actions. In this situation, therefore, *uncertainty refers to the lack of knowledge that would enable outcomes to be predicted.* Understood this way, uncertainty has all of the characteristics of the ordinary concept described above. Risk and uncertainty are not mutually exclusive. For instance, to take a risk is often to pursue an option whose consequences are more difficult to predict, and thus more uncertain, than available alternatives. Similarly, ignorance of some possible outcomes is a factor that makes prediction more difficult and, hence, which intensifies uncertainty. Finally, knowledge of probability is not always sufficient for prediction, as the example of the roulette wheel illustrates.

But formulations of PP often refer not merely to uncertainty but to *scientific* uncertainty. In the next section, I make the case that inability to predict rather than ignorance of probability is the key concept for explicating scientific uncertainty in relation to PP.

### *5.2.2 Scientific uncertainty and predictive validity*

In this section, I propose that scientific uncertainty be understood as the lack of a model whose predictive validity for the task in question is empirically well confirmed. Since empirical confirmation is a matter of degree, this definition entails that there is no sharp dividing line between cases in which scientific uncertainty is present and those in which it is not. I approach this predictive conception of scientific uncertainty through a consideration of a related proposal by Aven (2011). I examine several difficulties confronting Aven's proposal (Cox 2011) and then present my own. Finally, I discuss the rationale for making prediction rather than knowledge of probability the central concept in a definition of scientific uncertainty used to explicate PP.

Aven's conception of scientific uncertainty centers on "the problem of establishing an accurate prediction model (cause–effect relationship)" (Aven 2011, p. 1515). A prediction model is defined as a function, or set of functions, from inputs to relevant outcomes used to characterize the risk. Following Aven's exposition (2011, p. 1522), consider an example in which the prediction model consists of a single equation $Z = G(X_1, \ldots, X_n)$, where $Z$ is the outcome and $X_1, \ldots, X_n$ are the set of inputs and G is the function. For instance, in a risk analysis concerning rear- versus front-facing child car seats, $Z$ might represent number of fatalities, and the inputs could represent such things as the severity of the traffic accident, whether the child was restrained in a front- or rear-facing car seat, type of car, etc. The inputs would be associated with a probability distribution in which probabilities are interpreted in an objective sense and estimated on the basis of statistical data. However, uncertainties about these objective probabilities may exist. Those uncertainties can then be represented by another distribution in which probabilities are given a Bayesian interpretation as representing degrees of confidence that are informed by relevant background information. Given this set-up, Aven proposes that scientific uncertainty is absent when the accuracy of the prediction model $Z = G(X_1, \ldots, X_n)$ is well established empirically and the uncertainty of $X_1, \ldots, X_n$ is "small" (*ibid.*). Scientific uncertainty, then, arises if and only if one of these two conditions is not satisfied: either the uncertainty of $X_1, \ldots, X_n$ is "large" or the accuracy of the prediction model cannot be established (*ibid.*).

One immediate concern about Aven's account of scientific uncertainty is that it appears to be circular, as the term "uncertainty" itself appears in the definition. Aven's use of "uncertainty" within his definition appears to track the decision-theoretic conception. That is, the inputs to the predictive function are "uncertain" when probabilities construed as frequencies cannot be established for them. This interpretation of Aven's proposal is supported by two schematic examples he provides of scientific uncertainty (2011, p. 1523). In both cases, the frequency distribution of $X_1, \ldots, X_n$ cannot be estimated from statistical data, and, moreover, the variance of the Bayesian distribution over the possible frequency distributions is large. In other words, the frequency probability distribution over $X_1, \ldots, X_n$ is not known.

The main difference between Aven's proposal and the decision-theoretic definition, then, appears to be that Aven's does not treat risk and uncertainty as mutually exclusive. In his proposal, scientific *un*certainty is contrasted with scientific certainty, not risk (Aven 2011, p. 1522). In addition to this, Aven's proposal, unlike the decision-theoretic definition, explicitly links uncertainty to the inability to make accurate predictions. However, including the decision-theoretic definition of uncertainty as a subcomponent results in a proposal that is ultimately unclear. Is scientific uncertainty about lack of knowledge of probabilities interpreted as frequencies or the inability to predict? Since these two things do not always coincide, the answer to this question matters.

In a commentary on Aven's proposed definition of scientific uncertainty, Louis Cox (2011) raises an objection that is closely related to this point. Cox points out that the conditions required for scientific certainty in Aven's model might be satisfied without accurate prediction being possible (2011, pp. 1531–2). Situations of this sort occur for models that exhibit sensitive dependence on initial conditions (SDIC), that is, wherein small differences in input generate large differences in output (Werndl 2009, p. 203). In such cases, even if the prediction model is known and very little uncertainty exists about the values of the input variables, accurate predictions may not be possible, a point Cox illustrates by means of a logistic map model of an epidemic. Moreover, Cox points out that the predictive accuracy of SDIC models depends on the predictive task. In general, the further in the future the prediction, the less accurate the predictions will be. Of course, the difficulty of small errors of input being amplified into large errors of output is even more severe in the normal situation in which the model itself is not exactly correct (see Frigg, Smith, and Stainforth 2013).

I propose, then, to define *scientific uncertainty* about outcomes of actions as the absence of a model whose predictive validity regarding those outcomes is well confirmed empirically. This definition is similar to the predictive explication of the ordinary concept of uncertainty discussed in the previous section. It differs from that intuitive concept only insofar as specifying the means by which predictions are normally made in science, that is, on the basis of an empirically confirmed model of the phenomenon. This definition, therefore, aims to clarify the scientific part of scientific uncertainty. Elaborating the definition requires specifying the meanings of its two key components: "predictive validity" and "empirically well confirmed."

A predictive model is said to be valid if it is both accurate and precise. A model is accurate when its predictions are, on average, close to the actual value being predicted. A number of quantitative measures of predictive accuracy exist, such as the mean absolute error and the mean-squared error. Accuracy is standardly contrasted with precision, which refers to the tendency of predictions to be tightly clustered around a particular value. Precision in this sense is possible without accuracy, since predictions could be clustered around a value that is far removed from the true one. Moreover, accuracy is possible without precision. For example, accuracy as measured by mean absolute or squared error would be zero if predictions were symmetrically distributed in a huge area centered on the actual value to be predicted. Precision in addition to accuracy is important for prediction tasks, such as forecasting the quantity of sea level rise in the next century. In such a case, a model whose predictions of sea level increases are accurate on average but very imprecise will not be particularly helpful. The fact that errors would cancel out in the long run is cold comfort if the forecasts for this century go disastrously wrong (see Hansson 1993, pp. 22–5).

How, then, to empirically assess the predictive validity of a model for a task? I suggest the following list of considerations:

1. *Predictive success*: Does the model have a strong record of validity on relevantly similar prediction tasks?
2. *Model plausibility*: Is the structure of the model is grounded in established scientific knowledge, such as physical laws or well-understood biological processes? Is there strong empirical evidence for key parameter values and inputs of the model?
3. *Underdetermination*: Do other models exist that score reasonably well with respect to considerations 1 and 2 but which make significantly different predictions?

*Yes*, *yes*, and *no* answers to these questions, respectively, are sufficient for the predictive validity of a model to be empirically well confirmed. However, not all of the three criteria are necessary in every case. For example, the predictive validity of a model might be strongly confirmed empirically without being grounded in any well-established theory (Cox 2011, p. 1532). Of course, the answers to questions 1 through 3 are not strictly categorical but come in degrees, so assessments of empirical confirmation of predictive validity are not all or nothing. Scientific uncertainty, therefore, may exist to a greater or lesser extent in one context than another. In addition, answers to any one of the questions may be difficult to decide. In the case of underdetermination, for example, it may be difficult to rule out the possibility of some unconceived yet equally plausible and predictively successful alternative. Finally, judging a model's predictive validity for a particular task to be well confirmed should not be confused with inferring that the model is true (Oreskes, Shrader-Frechette, and Belitz 1994).

Consider three criteria listed above in relation to general circulation models (GCMs) used to predict the future course of climate change. Such models are grounded in well-established physical theories, but the translation from the basic physical principles that govern the atmosphere to a GCM is a treacherous one for two main reasons. First, numerous simplifications and idealizations must be made for the sake of computational tractability. Second, some important atmospheric processes, such as cloud formation, are not yet fully understood. Thus, criteria 1 and 3 are important for assessing the predictive validity of GCMs. A number of studies have in fact been published assessing the validity of GCM predictions made in the 1990s (Allen, Mitchell, and Stott 2013; Fildes and Kourentzes

2011; Fu *et al.* 2013; Rahmstorf, Foster, and Cazenave 2012).[1] Although preliminary, these studies suggest that GCMs used by the IPCC have done a good job predicting average global temperature increases in the past decade, but have tended to significantly underestimate sea level rise (Rahmstorf, Foster, and Cazenave 2012). Some of these studies also compare GCMs to alternative models in an apparent attempt to address issue 3 (Allen, Mitchell, and Stott 2013; Fildes and Kourentzes 2011). However, this aspect of assessing the predictive validity of GCMs remains very preliminary, in part because the rationale for choices of comparison models is unclear. At present, then, there appears to be moderate confirmation of the validity of decadal predictions of climate models concerning quantitative increases in global average temperature. However, scientific uncertainty is defined in relation to a prediction task. The predictive validity of GCMs for the long term is less well confirmed than for the short term. In addition, all of the studies cited above except for one (Fu *et al.* 2013) assessed the predictive validity of GCMs for global rather than regional effects of climate change. Hence, scientific uncertainty is greater with regard to these two latter types of predictions (Mearns 2010; Oreskes, Stainforth, and Smith 2010).

The analysis of scientific uncertainty proposed here is similar to the decision-theoretic definition in some respects. A true probability model, where those probabilities represent a stochastic process, is a predictively valid model for some, although not all, predictive tasks. For example, knowing that the half-life of carbon-14 is 5,730 years enables predictions of some things (e.g., approximately what proportion of a kilogram of carbon-14 atoms will decay in a 10,000 year period?) but not others (e.g., will this particular carbon-14 atom decay in a time period equal to its half-life?). In addition, providing evidence to support the predictive validity of a model typically requires analysis of statistical data. But in my proposal that statistical analysis is not of a naive empiricist sort wherein one attempts to estimate the relative frequency of an event (e.g., a three-meter rise in sea level by 2100) by counting how many times it occurred in a sample of similar cases. Instead, the statistical data are used to assess the predictive validity of a model, or models, in a range of predictive tasks that are relevantly similar to the one in question (e.g., on a similar timescale and topic). Furthermore, since my proposal does not define scientific uncertainty as ignorance of probabilities, it avoids ambiguities related to

[1] Validation studies of GCMs have also been performed with respect to paleoclimate data (Price *et al.* 1997).

whether probabilities are interpreted as degrees of confidence or objective chances. No assumptions are made about the interpretation of probabilities used in predictive models. According to my proposal, the important question is not about the interpretation of the probabilities but about the predictive validity of models used to infer the consequences of actions.

This definition of scientific uncertainty has a number of other advantages over the decision-theoretic definition. First, it tracks the characteristics of the ordinary conception of uncertainty described in the previous section. Unlike the decision-theoretic definition, mine does not treat risk and uncertainty as mutually exclusive. And unlike the decision-theoretic definition, mine does not make knowledge of all possible consequences of available options a necessary condition of scientific uncertainty. Omitting significant potential consequences is one reason a predictive model may fail to be valid. In addition, according to my proposal, knowledge of probabilities of outcomes is not sufficient to eliminate scientific uncertainty. According to the definition proposed here, scientific uncertainty might be quantifiable or unquantifiable by probabilistic means, where the later would entail an inability to assign probabilities to outcomes in a non-arbitrary way. The relevance of this to environmental decisions is illustrated by examples of chaotic models, such as the logistic map mentioned above. Imagine an ideal case in which one knows the true model that generates outputs from inputs and, moreover, that the objective chance distribution over the input variables is also exactly known. However, since the model is chaotic, the current distribution of input variables becomes progressively less informative about future outcomes, making it impossible to predict the distant future with any degree of precision (Werndl 2009, pp. 214–16). Since predictive validity is steadily lost as the model's forecasts are projected further into the future, my proposal entails that whether and to what degree scientific uncertainty is present depends on the timescale of the prediction. However, the decision-theoretic definition asserts that uncertainty is absent whenever probabilities are known and does not require that those probabilities be informative for the predictive task in question. Hence, the decision-theoretic definition provides no basis for the obvious judgment in this case that long-term predictions are more uncertain than near-term ones.

Finally, there is a sounder rationale for the definition of scientific uncertainty proposed here in relation to PP. That rationale is founded on the premise that PP is often applied to problems in which the ability to validly predict consequences of decisions – especially consequences that may be seriously harmful – is limited. Almost every important environmental and public health policy issue has this characteristic, as is illustrated by the case of climate change and the examples surveyed in Chapter 4. Moreover, when the ability to predict is limited, it is unwise to base significant policy decisions on what is sometimes termed a "predict and act" methodology (Mitchell 2009, chapter 5; Popper, Lempert, and Bankes 2005). In such an approach, one attempts to predict the consequences of available options and select the one whose consequences are, all things considered, the best. Cost–benefit analysis is, naturally, a methodology of this sort. Thus, the interpretation of scientific uncertainty proposed here supports the critique of attempted cost–benefit analyses of climate change mitigation given in section 2.3.

In contrast, the decision-theoretic definition of uncertainty seems to be motivated by the question of when maximizing expected utility or some other rule should be the criterion of rational decision. Since expected utility of an action is a probability weighted average of utilities, the reasoning is that if probabilities are unknown, then expected utility cannot be computed and hence some other decision rule, such as maximin, should be applied. Since the maximin rule is one commonly proposed basis for interpreting PP (see section 3.3.2), the definition of scientific uncertainty as ignorance of probabilities would seem to neatly demarcate PP's scope of application. However, this line of reasoning is problematic. First, it is invalid because probabilities alone are not sufficient for applying maximizing expected utility as a decision rule; a utility function is also needed. Thus, even if one grants the underlying assumption that PP picks up where utility maximization leaves off, it does not follow that ignorance of probabilities is a necessary condition for applying PP. Second, the decision-theoretic definition ties uncertainty specifically to a frequency concept of probability. But expected utility can be calculated with any type of probabilities, *including personal probabilities*. Indeed, that is the way in which expected utility is understood in Bayesian decision theory (see Jeffrey 2004). Thus, ignorance of probabilities interpreted as frequencies, or objective chances more generally, does not preclude applications of expected utility reasoning in either theory or practice. A defender of the standard definition of uncertainty might reply that responsible decision making requires that probabilities be relied upon only if they are empirically well founded in some sense. I sympathize with this response but suggest that working out its details leads to the concept of scientific uncertainty I propose here.

## 5.3 Probability and the limits of precaution

The decision-theoretic risk versus uncertainty distinction is sometimes used to demarcate the scope of application of PP: risk is said to be the province of quantitative approaches such as cost–benefit analysis, while uncertainty is the domain of PP (Dana 2009; Resnik 2004; Whiteside 2006). But others reject this division of labor and argue against treating ignorance of probability as a necessary condition for applying PP (Munthe 2011; Thalos 2009; Trouwborst 2006). Rejecting the decision-theoretic risk versus uncertainty dichotomy, as argued for above, naturally lends support to the latter of these two positions. But this perspective raises new questions of its own. In particular, if ignorance of probabilities is not a necessary condition for applying PP, then what is the relationship between PP and quantitative decision-making approaches? Moreover, some critics argue that to the extent that PP is defensible it is merely a version of existing quantitative approaches, such as cost–benefit analysis (Goklany 2001; Marchant 2001a; Posner 2004, p. 140; Soule 2004; Sunstein 2001, pp. 104–5). So granting that PP can be used in conjunction with quantitative approaches to risk raises important questions about how PP is distinct from those approaches. In section 5.3.1, I consider the relationship between PP and risk analysis, while in section 5.3.2, I examine the relationship between PP and cost–benefit analysis.

### *5.3.1 Precaution and risk analysis*

Risk analysis is often characterized as aiming to provide quantitative descriptions of risks, where risk is normally defined as the probability of an adverse outcome multiplied by a measure of the severity of that outcome, such as number of deaths (Aven 2003, 2008; Bedford and Cooke 2001).[2] However, the term "risk analysis" is sometimes construed more broadly to encompass normative approaches for deciding which risks should be run (Hansson 1993). For the purposes of this section, I will stick with the narrower definition of "risk analysis," while concerns related to risk analysis in the broader sense arise elsewhere in a variety of contexts.[3] This section examines the relationship between risk analysis – understood as providing

[2] This definition of risk is not universally adopted in risk analysis. For example, Aven (2008, p. 19) defines risk in terms of the uncertainty and severity of harm, where uncertainty is not always represented probabilistically.

[3] For instance, see sections 2.3, 5.3.2, Chapter 6, and section 8.4.

quantitative estimates of risks – and PP, making the case that the two are compatible and can be jointly applied.

Given the decision-theoretic risk versus uncertainty distinction and the premise that PP is applicable only to decisions under uncertainty, there would be almost no overlap between PP and risk analysis. However, as noted in section 5.2.1, the decision-theoretic definition does not accord with the actual practice of risk analysis, wherein assessments of uncertainty are a standard procedure. In addition, several authors argue against the presumption that ignorance of probability is a *sine qua non* of applications of PP. Trouwborst makes an argument based on the actual use of PP, stating that "not a single category of uncertainty can be considered as outside the reach of the precautionary principle in customary international law" (2006, p. 91). Thalos (2009) and Munthe (2011) propose arguments of a more conceptual nature. For her part, Thalos describes several aspects of precaution (see section 3.2) and points out that some of these may be relevant even when probabilities of outcomes are known (2009, p. 43). As discussed in section 4.4.1, Munthe interprets PP in terms of a duty to avoid irresponsible impositions of risks, and he insists that knowledge of probabilities of risks in no way eliminates this duty (2011, p. 48). The account of scientific uncertainty proposed above lends further support to such arguments. It rejects the contrast between risk and uncertainty, and it proposes a definition of uncertainty that is focused on the lack of a well-confirmed predictive model rather than ignorance of probabilities. And since it allows that scientific uncertainty may be quantifiable or unquantifiable, the definition of scientific uncertainty proposed here does not suggest that the domain of PP and that of risk analysis are mutually exclusive.

One way PP and risk analysis could be combined is suggested by alternatives assessment, which is often proposed as a means of implementing PP (O'Brien 2000; Raffensperger and Tickner 1999; Steinemann 2001; Tickner 2002; Tickner and Geiser 2004). Alternatives assessment emphasizes the systematic search for feasible and safer alternatives to materials, procedures, or activities currently in use. Rather than focusing time and resources on attempts to quantify toxic effects as precisely as possible and to decide whether actual exposures rise above levels deemed to be "safe," alternatives assessment recommends that greater effort be dedicated to finding solutions that avoid reliance on toxic materials or hazardous activities in the first place. Alternatives assessment also addresses issues relating to scientific uncertainty by posing questions that may demand less in the way of precise quantitative answers. Instead of needing to precisely quantify a risk associated with some activity, the problem is to decide whether one alternative is riskier than another. When the differences between the adverse effects of alternatives is large, then the preference for one over the other may be clear even if precise quantitative estimates are highly uncertain. The Massachusetts Toxics Use Reduction Act (TURA), originally passed in 1989 and updated in 2006, is often cited as an example of a successful implementation of the ideals of alternative assessment (O'Brien 2000; Tickner and Geiser 2004). TURA sets out a list of chemicals of concern and requires companies who use more than specified quantities of these chemicals to submit biannual assessments of their use of these chemicals and plans for how to reduce or eliminate those usages. For the period 1990 to 2005, TURA is credited with a 40% reduction in the use of toxic chemicals in Massachusetts, and in a survey of Massachusetts companies subject to TURA, 41% reported financial savings as a result of toxic use reduction measures prompted by the act (Massey 2009, pp. 16, 42).

Risk analysis is relevant to alternatives assessment in at least two ways. First, risk analysis may be important for deciding which materials or procedures should be targets of alternatives assessment. Second, risk analysis may be useful for assessing whether alternative materials or procedures are indeed less risky than what they would replace. The development of alternatives to traditional tin/lead solder in electronics illustrates this last issue (Black 2005). A number of alternatives to tin/lead solder exist and are now in wide use, due in part to a ban on lead/tin solder implemented by the European Union in 2006. Mandating the substitution of an alternative for tin/lead solder promotes public health only if these alternatives are indeed less hazardous. Discussing potential adverse effects of each alternative is, moreover, a basic component of alternatives assessment (O'Brien 2000, p. 147).

It may seem surprising to present alternatives assessment as an illustration of how PP and risk analysis can be integrated, since advocates of alternatives assessment sometimes present it in opposition to quantitative risk analysis (O'Brien 2000; Tickner and Geiser 2004). For example, the title of chapter 1 of Mary O'Brien's book, *Making Better Environmental Decisions*, is "Goal: Replace Risk Assessment with Alternatives Assessment" (2000, p. 3). O'Brien then proceeds to take the profession of risk analysis to task for being in the pocket of the industries whose products generate the risks they analyze and for systematically failing to consider a sufficiently broad range of alternatives. But there is no conflict between alternatives assessment and risk analysis in principle. There is no reason why efforts to quantify risks must be inimical to efforts to find better alternatives.

On the contrary, the two are – or in an ideal world, *would be* – mutually supporting. So, I think objections to risk analysis found in the literature on alternatives assessment should be regarded as criticisms of common practices rather than an indictment of endeavors to quantify risk per se. Moreover, despite her strong rhetoric, some of O'Brien's own statements suggest that she too views risk analysis as potentially working in concert with alternatives assessment. For instance, she writes that risk assessors are one of her target audiences "because many of them can help create or insist upon opportunities to present their assessments of risks within larger alternatives assessments" (2000, p. xiii).

Other means by which elements of PP could be integrated with risk analysis could be discussed, such as an expanded role for public or stakeholder participation in the assessment of risks (Whiteside 2006). However, I hope that what has been said thus far is sufficient to illustrate how quantitative estimates of risks can be relevant to applications of PP. In the remainder of this section, I turn to the charge that, if coherent, PP is ultimately no different from quantitative approaches to risk with which it is typically contrasted, such as risk analysis or cost–benefit analysis (Goklany 2001; Marchant 2001a; Posner 2004, p. 140; Soule 2004; Sunstein 2001, pp. 104–5).

The next section discusses cost–benefit analysis and its relation to PP. Here I describe two important features of PP that distinguish it from risk analysis. First, risk analysis is primarily a descriptive endeavor: it aims to provide quantitative characterizations of risks, not to specify a rule for deciding which risks should be run and which should not. That differs from PP, which does function as a decision rule. As discussed in Chapter 2, PP provides a qualitative structure for decision making in such circumstances. A version of PP consisting of a knowledge condition, harm condition, and recommended precaution is identified that can be consistently applied to the case in question. Then among those policies that can be consistently recommended, the most efficient is sought. Second, PP is designed to be capable of guiding decisions even when no quantitative risk analysis exists. As was illustrated by the example of climate change mitigation discussed in section 2.4.2, PP can utilize qualitative rankings of knowledge conditions that may be justifiable even when scientific uncertainty is unquantifiable. The ability to function as a decision rule given only qualitative information is essential for meeting the demand of MPP that scientific uncertainty not be a reason for inaction in the face of serious threats.

However, a critic might object that without quantitative predictions about which outcomes are more or less likely, it is impossible to rank knowledge conditions and to compare the efficiency of alternative options (Turner 2013). I respond to this objection in two parts. First, knowledge insufficient for valid quantitative predictions on some important topic, such as the adverse impacts of climate change in the next century, may nevertheless allow for qualitative rankings of the plausibility of various future scenarios. As explained in further detail below, PP can utilize qualitative knowledge of this kind as a basis for decision making. Indeed, the ability to utilize qualitative information is one of the distinctive features of the precautionary principle. Second, the efficiency of options can be assessed by means of robust adaptive planning approaches that do not rely on quantitative predictions, as discussed in section 3.4.1. Robust adaptive planning is a decision framework explicitly designed for cases in which "predict and act" methodologies are inapplicable (Popper, Lempert, and Bankes 2005). Avoiding harmful side effects or other costs is an important objective in devising a robust policy, and hence robust adaptive planning is a plausible framework for interpreting efficiency. Since section 3.4.1 already discusses robust adaptive planning and its connection to efficiency, I focus on the first of these two responses here.

The concept of scientific uncertainty defended above is helpful for understanding how applications of PP can proceed in the absence of quantitative estimates of risks. Just as a model that is predictively valid for some task is not necessarily true, a model that provides an accurate representation of causal mechanisms of a phenomenon may fail to be predictively valid for many of its behaviors. Climate change is plausibly an illustration of this type of situation. Although a great deal is known about the physical mechanisms through which emissions of greenhouse gases affect the climate, quantitative long-term predictions of the course of climate change remain highly uncertain. And for reasons discussed in section 5.2.2 relating to chaotic systems, it is doubtful that scientific uncertainty about long-run predictions could be eliminated even if an almost perfectly accurate model of the global climate could be constructed. Climate change is not the only example of a scientific field in which substantial knowledge of mechanisms does not suffice for precise quantitative predictions of many important phenomena. For example, knowledge of natural selection and genetics does not enable predictions of where and when new antibiotic strains of bacteria will emerge. Nevertheless, this biological knowledge is obviously very relevant for decisions regarding responsible use of antibiotics.

Significant scientific uncertainty about future outcomes, then, can coexist with extensive knowledge of underlying mechanisms, which can provide a basis for qualitative rankings of knowledge conditions in an application of PP (cf. Resnik 2003, pp. 337–41). There is a difference between a purely hypothetical possibility (e.g., invasion by space aliens) and an outcome for which there is a scientifically established mechanism type whereby it could arise (e.g., the evolution of a strain of bacteria that is resistant to an antibiotic). And there is a difference again between a known mechanism type by which an outcome could arise and a specific mechanism observed to be in operation likely to lead to that outcome. Climate change fits the latter category: the mechanism type is known and evidence of its actual operation is accumulating from a wide variety of sources. However, such knowledge does not entail the existence of a model whose predictive validity regarding pressing issues has been empirically well confirmed. While there is a well-established correlation between GHG emissions and rising global temperatures and a physical mechanism to explain this correlation (Solomon *et al.* 2007), it is still very difficult to make informative and well-grounded predictions concerning many key outcomes related to climate change (Parry *et al.* 2007, p. 782).

### 5.3.2 *Value commensurability and cost–benefit analysis*

Although both are quantitative approaches, risk analysis and cost–benefit analysis differ in a number of important respects. Unlike risk analysis, cost–benefit analysis aims not merely to quantify risks but to provide a procedure for deciding which of several available options to choose. Moreover, while risk analysis typically quantifies risks in their natural units (e.g., the number of fatalities), cost–benefit analysis converts all costs and benefits to a common monetary scale. In this section, I suggest that PP differs from cost–benefit analysis in not presuming value commensurability, and hence the two approaches are distinct even when scientific uncertainties are relatively minor.

Let's begin with a concise definition of the concept. A person's valuations over a set $S$ are commensurable just in case there is some unit of value $v$ (dollars, grams of gold, bushels of rice, or whatever) such that for any $s \in S$ there is some finite quantity $q$ of $v$ such that the person is indifferent between $s$ and $q$. For example, suppose the set $S$ consists of an apple, a pear, and a banana. Then my values are commensurable over this set if, for instance, I am indifferent between 50¢ and the apple, 20¢ and the pear, and 75¢ and the banana. These preferences would also be assumed to be relatively stable and to generate decisions that the person makes, such as choosing an apple over a pear or refusing to pay $2 for a banana.

Value commensurability is a standard assumption of environmental applications of cost–benefit analysis (Jaeger 2005, p. 11; Sunstein 2001, p. 111). Cost–benefit analysis aims to select among alternative options, or policies, by tallying up the expected costs and benefits of each and choosing the option with the best difference of the two. This procedure requires converting all costs and benefits to some common measure, which is normally assumed to be monetary. Value commensurability has been the target of a good deal of criticism (Ackerman 2008a; Ackerman and Heinzerling 2004; Anderson 1993). Critics charge that monetary valuations of impacts on human health or the environment are arbitrary and that they obscure considerations pertaining to rights and justice that would normally be considered essential (see Ackerman and Heinzerling 2004). In this section, I consider two possible ways in which value commensurability might be defended from such critiques. Value commensurability might be viewed as an empirical claim about the sorts of preferences that people *actually have*, or it might be defended from a *pragmatic* perspective as a helpful exercise for making trade-offs explicit. I argue that value commensurability is not defensible under either interpretation.

Consider value commensurability as an empirical claim about the sorts of preferences that people actually have. A central argument for such an assumption is the observation that people frequently make decisions that involve trade-offs between supposedly incommensurable things, such as money and risk of a fatal accident. For example, people sometimes accept work that carries a higher risk of death in exchange for higher pay. Thus, given a large sample of data concerning such decisions, the argument could continue, it would be possible to infer the quantity of money that people, on average, demand in order to be subjected to, say, a 1 in 10,000 of risk of fatality. These inferred preferences, then, could be used to predict behavior – for instance, of how much consumers are willing to pay for reduced risks in a variety of other contexts. And they might also be used in a cost–benefit analysis of, say, a proposed regulation aimed at reducing air pollution.

A number of objections could be raised to the procedure sketched in the above paragraph. One might question why actual decisions some people make in one context – decisions that may not have been made with adequate information or with a decent range of options – should be normative for decisions affecting a possibly distinct group of people in a different context (see Ackerman and Heinzerling 2004, pp. 75–8; Anderson 1993, pp. 195–203; Kysar 2010, pp. 111–12). Another objection would be that estimates of the wage premium associated with increased workplace risk span an extremely wide range of figures, making the choice of any particular number arbitrary (Ackerman and Heinzerling 2004, p. 82; Anderson 1993, p. 196). However, my concern here is with the assumption that actual decisions involving trade-offs are generated by a fairly stable and context-independent preference structure (Sagoff 2008; Starmer 1996). This assumption is crucial to the approach, for otherwise there is no reason to suppose that trade-offs would be made in a similar manner in the future or in other circumstances. Likewise, there would be no grounds for using those inferred preferences as a basis for evaluating decisions made in other contexts. But the inference from observed trade-offs to underlying stable and value commensurable preferences is highly underdetermined by the data. Rather than being stable and context-independent, preferences regarding trade-offs may be shaped by a variety of factors specific to the context in which the decision is made. As Mark Sagoff puts the idea, "human beings adjust their choices in view of moral principles, opportunities, situations, occasions, and exhortations that may not reflect a preexisting stable preference schedule but may respond to qualities such as self-doubt, willingness-to-learn, decency, and social-mindedness" (2008, p. 78). In a similar vein, Chris Starmer proposes that decisions often involve a "rationalization strategy" that relies on "a bundle of decision heuristics, rather than a bundle of preferences" (1996, pp. 206–9).

The term "preference construction" is used in psychology, experimental economics, and market research to describe the processes through which preferences are generated in concrete circumstances. Preference construction appears to be common in decisions that involve trade-offs and decisions in which preferences must be expressed numerically (Lichenstein and Slovic 2006, p. 1), features that are typical of environmental policy decisions. Empirical research has produced a wealth of data supporting the idea that people often construct preferences in such situations rather than being guided by a set of preformed preferences that determine their choices (Hausman 1992; Lichenstein and Slovic 2006; Starmer 1996). For example, a variety of experiments demonstrate the effects of arbitrary anchors, such as the last four digits of a social security number, on individuals' stated willingness to pay (SWTP) for an environmental good, such as saving an endangered species (Kahneman, Ritov, and Schkade 2006). Yet if these decisions were generated by stable and value commensurable preferences, then the mention of an arbitrary number should have no effect on an individual's SWTP. In light of these and other results, Daniel Kahneman, Ilana Ritov, and David Schkade (2006) propose that a person's SWTP for an environmental good is primarily an expression of positive or negative attitudes – for example, about environmentalism or the social value of the market – rather than reflections of stable preferences that assign specific dollar values to a variety of environmental goods. If this is right, then there can be no hope of a scientifically well-founded method for uncovering the monetary value people assign to environmental goods, for the simple reason that people do not have the stable and commensurable preference rankings that such an exercise presupposes. As Kahneman, Ritov, and Schkade put it: "If consumer theory does not capture the nature of people's value for environmental goods, there can be no more hope of measuring the economic value of the whooping crane than there is of measuring the physical properties of the ether" (2006, p. 590).

In short, value commensurability as a descriptive hypothesis is an explanation of *how* people make decisions involving trade-offs, not merely the observation that people *do* make such trade-offs. Such trade-offs might be generated from stable preferences that assign monetary values to goods of all kinds, or the decisions may result from a combination of qualitative attitudes and features of the context in which the decision is made. And at present the evidence appears to favor the latter, preference construction, explanation. Turn, then, to the pragmatic defense of value commensurability. This defense agrees that value commensurability is not accurate as a psychological hypothesis but insists that it is nevertheless indispensable for reasoned decisions that involve balancing costs and benefits. Sunstein states this position as follows:

> Cost–benefit analysis should not be seen as embodying a reductionist account of the good, and much less as a suggestion that everything is simply a "commodity" for human use. It is best taken as a pragmatic instrument, agnostic on the deep issues and designed to assist people in making complex judgments where multiple goods are involved. (2001, p. 124)

Sunstein claims that requiring all goods at stake in an environmental policy decision to be converted to dollar values has the pragmatic value of forcing an explicit discussion of trade-offs (Sunstein 2001, pp. 111, 123–4). Moreover, this method might be thought to facilitate consistency in valuations: if the elimination of a 1 in 10,000 risk of fatality is judged to be worth a certain quantity of money in one context, then it should be assigned the same dollar amount in other decisions unless some justification for a difference can be provided. According to the pragmatic defense, then, value commensurability is simply a device for imparting greater transparency and rigor to reasoning underlying decisions involving hard trade-offs.

However, the pragmatic defense of value commensurability is problematic because it recommends adopting an assumption that it admits is significantly false. Basing an analysis on a false assumption is sensible only if there is some reason to think that the assumption, though false, is a sufficiently good approximation for its intended purposes. Yet Sunstein provides no reasons of this kind. And without such reasons, treating value commensurability as a basic premise may just as well be supposed to *distort* decision making as to clarify it (Bantz 1982, pp. 234–5). Most apparently, assuming value commensurability when it is substantially false would be likely to result in arbitrary numerical estimates of people's willingness to pay for reductions of risks or environmental benefits having a disproportionate influence on the outcomes of decisions. Another distortion arising from assuming value commensurability when it is false has to do with the choice of common unit. To see the idea, note that if value commensurability were true, then any number of units of value could be chosen as the common one – money, gold, grain, potatoes, or what have you. If every valued entity had a dollar value, then it would also have a corresponding value in gold, grain, and potatoes. The choice of unit, then, would be purely a matter of convenience. However, *if value commensurability is significantly false but assumed to be true in an analysis, then the choice of common unit is no mere matter of convenience but is likely to distort reasoning in systematic ways.* For if value incommensurability is false, then valuations of distinct types of entities or relationships may be subject to divergent sets of norms (Anderson 1993). Thus, falsely assuming value commensurability can distort reasoning when norms appropriate for the presumed common unit of value are not accepted for other types of valued things. For instance, treating money as the common unit of value will have the effect of substituting norms of financial reasoning for principles of justice and morality in ways that many would find highly objectionable. Examples of this sort of situation abound in critiques of cost–benefit analysis, as illustrated by critiques of economic analyses of climate change that assign a higher dollar value to the lives of individuals from wealthier nations (Ackerman and Heinzerling 2004, pp. 73–4). While such valuations may have an economic rationale in terms of willingness to pay, they are in stark conflict with basic moral principles concerning human equality. The case of future discounting, discussed in Chapter 6, provides another illustration of distortions that can ensue from presuming that norms pertinent to financial decisions should ground decisions of all sorts.

Consequently, if value commensurability is assumed in analysis for pragmatic reasons, the implications of its falsehood should be clearly borne in mind so as to avoid distorted reasoning. In particular, it should be explicitly recognized that if value commensurability is false, then the choice of common unit can have the effect of extending norms pertinent to that unit beyond their normal sphere of influence. Furthermore, if transparency of reasoning involving trade-offs is a desired goal, then alternative means of promoting it that are less prone to generating such distortions should be sought.[4]

The above discussion, then, highlights a substantial conflict between PP and cost–benefit analysis. Moreover, it is a conflict that is not primarily connected to knowledge of probabilities of outcomes, although PP and cost–benefit analysis can butt heads on that issue as well (see section 2.3). A false presumption of value commensurability with money as the common unit of value will have the effect of exporting norms of financial reasoning to contexts in which they would not normally be thought to be appropriate. One result of this may be to promote norms of decision making that undermine efforts to enact precautions in the face of serious environmental threats.[5]

## 5.4 Conclusions

The central contribution of this chapter is a definition of scientific uncertainty as the absence of a model whose predictive validity for the task in question has been well confirmed empirically. This proposal has a number of advantages over the commonly cited decision-theoretic definition of uncertainty. It tracks the intuitive concept of uncertainty more faithfully and is thereby less likely to produce confusion and miscommunication. In addition, there is a sounder rationale for its connection to PP, namely that precaution is often needed in circumstances wherein valid predictions of outcomes of actions may not be forthcoming. Finally, the new definition of scientific uncertainty results in a better understanding of the relationship between PP and quantitative methods used to assess environmental policy

[4] See Satterfield *et al.* (2013) for a discussion of some of the possible approaches.

[5] Although I do not explore the matter here, it is an interesting question whether cost–benefit analysis could be modified to be compatible with PP (see Revesz and Livermore 2008).

decisions, particularly risk analysis and cost–benefit analysis. One important result is that PP and cost–benefit analysis differ on how valuations of consequences should be carried out and hence could diverge even in an ideal case in which scientific uncertainty was entirely absent. This point comes front and center in the next chapter.

CHAPTER 6

# *Counting the future*

> Imagine finding out that you, having just reached your twenty-first birthday, must soon die of cancer because one evening Cleopatra wanted an extra helping of dessert.
>
> Tyler Cowen and Derek Parfit, "Against the Social Discount Rate," p. 145

## 6.1 How to count the future?

Decisions about long-term problems inevitably raise questions about how the interests of the present should be balanced against those of the future. Should present and future generations be treated impartially, or should present concerns be weighted more heavily than later ones? In economics, this topic is discussed under the heading of *discounting*. Whether one chooses to discount the future, and at what rate, can obviously have an enormous impact on policy decisions regarding environmental issues that play out over a period of decades or, in some cases, centuries. Moreover, discounting is highly relevant to PP, as precautions typically involve incurring some present cost or inconvenience for the sake of mitigating some future risk. In this chapter, then, I examine the relationship between discounting and PP.

Debates on discounting are often divided into two main approaches, descriptive and prescriptive (Arrow 1999; Frisch 2013; Nordhaus 2007). According to the descriptive approach, the future should be discounted by a factor that reflects the rate of return on investment, which is regarded as an indicator of the extent to which people actually discount the future in comparison to the present. The descriptive approach seems to be motivated by the anti-paternalistic conviction that it is not the job of economics to tell people what values they should have. Instead, the thought is that the role of economic advice is to assist people in making decisions that are rational given their preferences. Consequently, if people systematically discount the future, then this preference should be reflected in assessments of alternative environmental policies. In contrast, the prescriptive approach insists that judgments about whether to discount the future – and if so, at what rate – inevitably rest on premises concerning justice and fairness. Since there appears to be no moral basis for weighing the interests of people at one time more highly than those of people at other times, advocates of the prescriptive approach generally argue that the interests of all generations should be treated impartially (Broome 1992; Cowen and Parfit 1992; Gardiner 2011; Stern 2007). I refer to this position as *intergenerational impartiality*. Intergenerational impartiality does not prohibit discounting in all its guises and, hence, does not mean that the overall discount rate must be zero. Instead, it eliminates what is known as the "pure time preference" (Price 1993, chapter 8) or the "pure discount rate" (Broome 1992, p. 92), that is, *discounting the future simply because it is the future.* As discussed in section 6.2, the standard argument for intergenerational impartiality appeals to a principle of *universalizability*, according to which moral principles must apply uniformly to all persons, places, and times.

The contrast between descriptive and prescriptive approaches, then, frames the debate about discounting in terms of a choice between norms of instrumental rationality and principles of justice. In this chapter, I argue that this framing of the issue is problematic in two respects. First, as discussed in section 6.3, agent-relative ethics challenges universalizability on moral grounds, thereby calling into question the standard argument for intergenerational impartiality. Second, I argue in section 6.4 that instrumental concerns about how to effectively implement what I call *sequential plans* lead to an argument for intergenerational impartiality. A sequential plan consists of a series of steps enacted over an extended period time, such as a carbon tax ramp of the sort discussed in section 2.4.2, wherein a tax on the carbon content of fuels is implemented and gradually increased. A common feature of sequential plans is that those who frame the plan and take its first steps do not have the power to force future actors to follow the plan's directives. Consequently, a sequential plan is unlikely to be effectively implemented unless it is *sequentially justifiable*, that is, unless the plan can be rationally justified to actors at every step in the sequence. I suggest that sequential justifiability leads to an argument against the pure time preference and in favor of intergenerational impartiality. For actors at later times are very unlikely to share the judgment that the past should be valued more highly than their own present. Nor is there any plausible reason that could be given to these actors to adopt value judgments that discriminate against them in this way. Hence, future actors would have no reason to follow through on a plan that subjects them to a disproportionate share of burdens on the basis of a pure time preference applied at an earlier date.

The above argument for intergenerational impartiality, therefore, aims to establish the following conditional: *if* you are committed to implementing effective sequential plans to address environmental problems, *then* you should assume intergenerational impartiality. I think it is clear that anyone who accepts PP should share the commitment specified in the antecedent of this conditional and thus that intergenerational impartiality is an important component of PP. Moreover, it is not difficult to argue that a moral obligation exists to implement sequential plans on issues such as climate change or stratospheric ozone depletion. Since the argument for this obligation need not presume universalizability, the case made here for intergenerational impartiality deftly sidesteps the challenge posed by agent-relative ethics. Finally, it shows that the descriptive approach to discounting is flawed on instrumental grounds. Given that people's preferences will change across time – for instance, every generation will regard its own present as more important than other times – grounding long-term policies on actual preferences of the present is unwise. Instead, effective sequential plans should be grounded in reasons that can be rationally accepted by actors at each step, and that in turn leads to intergenerational impartiality.

## 6.2 Intergenerational impartiality and discounting

John Broome states intergenerational impartiality as follows:

> A universal point of view must be impartial about time, and impartiality about time means that no time can count differently from any other. In overall good, judged from a universal point of view, good at one time cannot count differently from good at another. Nor can the good of a person born at one time count differently from the good of a person born at another. (1992, p. 92)

The "universal point of view" mentioned in this passage refers to the Kantian idea that moral prescriptions must be universalizable, an idea also taken up in the utilitarian tradition. R. M. Hare defines universalizability as follows: "Universalizability . . . comes to this, that if we make different moral judgements about situations which we admit to be identical in their universal descriptive properties, we contradict ourselves" (Hare 1981, p. 21). A universal descriptive property is a property that makes no reference to any particular person, time, or place. To disvalue the interests of people living at one time in comparison to those at another time *solely because of the difference in times*, therefore, would be impermissible given universalizability. Universalizability is intended to express a basic aspect of fairness, since it demands that a moral right or obligation pertaining to one person must also hold for every similarly situated person in any other time or place. Intergenerational impartiality extends this idea from individual persons to generations. According to the notion of sustainable development, for instance, each generation has a right to use resources to fulfill its own needs and an obligation to refrain from activities that undermine the capacity of future generations to do likewise (Norton 2005). Thus, the standard argument for intergenerational impartiality is that it is a straightforward consequence of a basic understanding of fairness (Broome 1992, pp. 92–108; Gardiner 2011, pp. 265–97; Stern 2007, p. 35).[1]

Let us turn, then, to the relationship between intergenerational impartiality and the economic concept of discounting. If $x$ is some benefit or cost, then discounting is usually defined mathematically by the following equation:

$$PV(x_t) = x_t/(1+r)^t.$$

In this equation, $t$ indicates units of time, $PV(x_t)$ is the present value of obtaining $x$ at time $t$, and $r$ the discount rate. For example, if years are the unit of time and $r$ is .05, then $1,000 five years from now is worth about $784 today. Discounting according to the above formula is known as *exponential* discounting and is the mathematical form of discounting normally used in cost–benefit analysis. For decisions involving long-term consequences, the choice of $r$ can clearly have substantial implications for which options are preferable. In general, a higher discount rate will minimize the importance of more temporally distant costs and benefits in comparison to near-term ones. The discount rate can arise in either private or public decisions (e.g., a person deciding between a fifteen- or thirty-year mortgage versus a government agency deciding whether to impose a stricter standard for ground-level ozone). It is common to use the term *social discount rate* to distinguish the latter case from the former and to highlight the fact that the two discount rates are not necessarily equal (Stern 2008, pp. 12–13). Since the social discount rate will almost always

1 Pearce (1998, p. 273) objects to Broome's argument for intergenerational impartiality by criticizing a particular version of utilitarianism endorsed by Broome. However, this misconstrues the argument for intergenerational impartiality. That argument rests on universalizability, which is compatible with a variety of ethical theories.

be the concern here, “discount rate” should be understood to mean “social discount rate” in what follows unless otherwise indicated.

Given its potential to impact long-term decisions, it is not surprising that the choice of discount rate is a hotly debated topic in the economics of climate change. Discussions of the discount rate are often framed in terms of the Ramsey equation, which arose from an analysis of the optimal savings rates of nations (Ramsey 1928). This equation is the following:

$$r = \rho + \eta g,$$

where $r$ is the discount rate as before, $\rho$ is the pure time preference, $\eta$ is the elasticity of marginal utility, and $g$ is the average rate of growth in consumption. The pure time preference, $\rho$, represents a devaluation of the future simply because it is the future. Thus, $\rho > 0$ indicates a preference for postponing a cost until later even if the adverse effect of the cost, when it happens, is the same. The growth rate, $g$, reflects the average rate of economic growth adjusted for inflation. In the context of climate change economics, the significance of $g$ for discounting is that if $g$ is positive, then future generations will be wealthier than current ones and, consequently, better able to pay for reductions in GHG emissions and adaptations to climate change. Finally, the elasticity of marginal utility, $\eta$, represents the relative curvature of the social utility function for consumption. It is common to assume a concave utility function, which represents both the declining marginal utility of wealth and aversion to risk. In this case, $\eta > 0$, which would mean that the greater $g$, the greater the discount rate $r$. In the context of climate change, this would imply a rationale for discounting, namely that future generations will be wealthier and hence should devote a greater proportion of their wealth to mitigating climate change than present generations. Ideally, the growth rate $g$ would be forecast on the basis of economic data. That would then leave $\rho$ and $\eta$ to be decided upon. And decisions about what values to assign can have major implications. For instance, in the economic analyses of climate change mitigation discussed in section 2.3, Stern sets the values of $\rho$ at 0.1 and $\eta$ at 1, while Nordhaus sets these parameters at 1.5 and 2, respectively. As a result, $r$ is much higher in Nordhaus's analysis than Stern's, and this difference is one important factor explaining their divergent recommendations.

Before discussing the implications of the Ramsey equation, it is important to note that the equation is derived from an idealized model of savings decisions. Ramsey describes this model as follows:

> ... our community goes on for ever without changing either in numbers or in its capacity for enjoyment or in its aversion to labour; that enjoyments and sacrifices at different times can be calculated independently and added; and that no new inventions or improvements in organisation are introduced save such as can be regarded as conditioned solely by the accumulation of wealth. (1928, p. 543)

The simplifying assumptions of this model should be borne in mind when considering the relevance of the Ramsey equation to real-world cases. In particular, rejecting the assumption of an unchanging utility function across time is central to the argument advanced below concerning sequential plans. Nevertheless, the Ramsey equation provides a helpful structure for discussing the concept of intergenerational impartiality.

In terms of the Ramsey equation, intergenerational impartiality asserts that the pure time preference, $\rho$, should equal zero. But intergenerational impartiality does not preclude discounting altogether. Instead, it insists that any discounting of one time with respect to another must be justified by some relevant difference between the two. By analogy, an impartial teacher is not required to give identical grades to all students on an exam, but fairness does demand that distinct grades be awarded only when some relevant difference exists. Such ideas are represented in a simple way by the Ramsey equation. Setting $\rho = 0$ does not entail that the overall discount rate is zero, since $\eta g$ may not be zero. If $\eta$ and $g$ are both positive, then intergenerational impartiality may be compatible with discounting on the grounds that future generations will be wealthier than the present. In relation to PP, this means that it *may* be acceptable to discount on the basis of a positive value of $g$ because increased wealth blunts the severity of many types of losses. For instance, future discounting may be reasonable if the costs in question are, say, losses of property due to extreme weather events. But the rationale for future discounting on this basis depends on the harm being mitigated by increased wealth, while the capacity of wealth to blunt the severity of harm may be limited by several factors. Some types of harms, such as the loss of human life, are not fully economically compensable. The primary harm of death is not mitigated by wealth – being wealthy does not somehow make your death less undesirable. However, wealth does mitigate some harmful side effects of death, such as impoverishment of dependents of the deceased. Furthermore, the mitigation of harms by wealth depends not only on the *total* wealth of a society but also on its *distribution*. If the gains in wealth are concentrated in a small minority, then increased average wealth may do little to mitigate environmental harms that fall more heavily upon the less well off (Broome 1992, pp. 67–8).

Finally, intergenerational impartiality is also compatible with an epistemic form of discounting. It is often the case that events further off into the distant future are more uncertain than those closer at hand. In PP, epistemic discounting would be reflected in the knowledge condition rather than in the assessment of how severe the harm would be if it occurred. When the potential harm satisfies only a very minimal knowledge condition, consistency will make it difficult to recommend any drastically costly precaution to prevent that harm (see section 2.4.1). In the Ramsey equation, uncertainty about the future could be considered in several ways, the most obvious being in connection with the rate of economic growth $g$. Since the discount rate depends on $g$, the Ramsey equation entails that, even given fixed values for $\rho$ and $\eta$, there is no unique discount rate when the rate of future economic growth is uncertain (as it actually is, of course). In a scenario in which $g$ is negative in the long run – for instance, due to adverse impacts of climate change – the value of $r$ could be negative (Stern 2007, p. 36). If the uncertainty regarding $g$ were quantifiable, one might construct a probability distribution over $g$, which would (again, given fixed values for $\rho$ and $\eta$) allow one to compute an expected value for $r$ (Weitzman 2007, p. 706). If the uncertainty regarding $g$ were unquantifiable, then numerical values for $r$ would be arbitrary.

Sometimes uncertainty about the future is folded into the Ramsey equation in one limited respect, namely through the addition of a factor representing the chance that the human species will go extinct (Dasgupta and Heal 1979; Pearce 1998, p. 272; Stern 2007, p. 53). For example, the Stern *Review on the Economics of Climate Change*, sets $\rho$ to 0.1 to reflect the possibility that the human species will cease to exist at some point in the future. However, I think it is better that the impact of uncertainty about the future be represented by uncertainty about $r$ (Broome 1992, p. 102).[2] Separating uncertainty from assessments of value is in keeping with the normal approach to expected utility, wherein utilities are a function of possible states of the world and probabilities represent uncertainty about which state is the actual one. Moreover, the possibility of human extinction is an epistemic reason for discounting, and not a matter of discounting the future simply because the future is judged less valuable than the present (Cowen and Parfit 1992, p. 147). In what follows, then, I will take $\rho$ to represent the pure time preference alone, not uncertainty about the continued existence of the human species in the distant future.

[2] Indeed, as noted above, the Ramsey equation is derived from a model in which it is assumed that the community exists forever (1928, p. 543).

## 6.3 Arguments for the pure time preference

In this section, I examine three interrelated arguments in defense of the pure time preference and objections to them that have been made by advocates of intergenerational impartiality:[3] the argument from excessive sacrifice, the argument from actual time preferences, and the argument from special relations. These arguments are connected insofar as they can all be construed as aspects of a challenge to universalizability, and hence to intergenerational impartiality, posed by an agent-relative perspective on ethics. Moreover, previous replies to the argument from special relations ultimately grant an ethical rationale for a pure time preference but point out that this time preference cannot be captured by an exponential discount rate. While important, this reply abandons intergenerational impartiality. One who wishes to defend intergenerational impartiality, therefore, must develop some further argument against a pure time preference in the context of intergenerational decisions.

### *6.3.1 The argument from excessive sacrifice*

Ramsey's own deeply ambivalent attitude toward intergenerational impartiality provides an intriguing backdrop for arguments about the pure time preference. In an oft-quoted passage[4] at the start of his (1928) essay, Ramsey strongly expresses his commitment to intergenerational impartiality: "it is assumed that we do not discount later enjoyments in comparison with earlier ones, a practice which is ethically indefensible and arises merely from the weakness of the imagination" (Ramsey 1928, p. 543). But a moment's reflection makes this statement very puzzling. If it is assumed that $\rho$ should obviously be zero, why propose an equation for the discount rate that includes $\rho$ as one of its terms? The remainder of the sentence just cited suggests the answer: "we shall, however, in Section II include such a rate of discount in some of our investigations" (Ramsey 1928, p. 543). In other words, despite his theoretical misgivings, Ramsey does in fact assume a value of $\rho > 0$ in developing his model in relation to actual economic behavior. He does this because setting $\rho = 0$ in his model leads to two

3 For reasons of space and thematic focus, I do not examine the full list of arguments on the topic of discounting. Those interested in arguments not considered here may see Broome (1992, chapter 3), Cowen and Parfit (1992), Dasgupta and Heal (1979), Gardiner (2011, chapter 8), Revesz and Livermore (2008), and Stern (2008).

4 For example, this passage is cited in Broome (1992, p. 96), Price (1993, p. 101), and Stern (2008, p. 15).

problematic consequences (Ramsey 1928, pp. 548–9). First, it results in rates of savings that are far greater than what individuals or states actually save and which, from an intuitive standpoint, seem unreasonably high. Second, it leads to the odd and implausible result that the rate of savings is independent of the rate of interest. Thus, Ramsey supposes $\rho > 0$ and shows that this results in lower savings rates and a functional dependence of savings on interest rates in his model.[5]

Ramsey's ambivalence is interesting because it is mirrored in the contrast between the stated opinions and actual practices of subsequent economists. It is not difficult to find strong statements against the pure time preference from prominent economists, such as this one from the Nobel laureate Robert Solow: "In social decision-making . . . there is no excuse for treating generations unequally, and the time horizon is, or should be, very long" (Solow 1974, p. 9).[6] Despite such statements, the decision to set the pure time preference to 0.1 in the Stern *Review on the Economics of Climate Change* (Stern 2007, p. 53) was extremely controversial. As Martin Weitzman puts it: "For most economists, a major problem with Stern's numbers is that people are not observed to behave as if they are operating with $\rho \approx 0$ and $\eta \approx 1$" (Weitzman 2007, pp. 708–9).[7] Most economists, Weitzman suggests, would prefer a value of $\rho$ closer to 2 (Weitzman 2007, p. 707). Moreover, two common arguments for the pure time preference parallel the reasons that drove Ramsey to suppose $\rho > 0$. Namely that a value of $\rho$ close to zero would entail (1) excessive present sacrifices for the benefit of the future and (2) the lack of an appropriate relationship between interest rates and discounting. In this section, I examine the first of these two issues.[8]

One of the most common arguments in favor of the pure time preference is that setting $\rho = 0$ would lead to taking drastically expensive actions now to bestow small benefits upon a huge number of future people (Arrow 1999; Nordhaus 2008, pp. 183–4; Pearce, Atkinson, and Mourato 2006, p. 185). The idea is that without a positive value for the pure time preference, accumulated small benefits or costs incurred by the potentially limitless number of future people would swamp almost all present

[5] Note an alternative course here that Ramsey does not consider: perhaps the problematic results are due to a failure of his model to adequately correspond to reality.

[6] See Price (1993, p. 101) and Stern (2008, p. 15) for additional references to similar statements.

[7] For consistency with the exposition of this chapter, I substitute $\rho$ for $\delta$, which Weitzman uses to denote the pure time preference see Weitzman (2007, p. 706).

[8] For critiques of the argument from interest rates, see Broome (1992, pp. 60–92) and Stern (2008).

concerns, resulting in a "dictatorship of the future" (Chichilnisky 1996). Thus, the argument from extreme sacrifice insists that a pure time preference is necessary to prevent responsibilities to the future from being transformed into unreasonable burdens on the present.

Derek Parfit (1984, pp. 484–5) objects to this argument on the grounds that it provides no basis for discounting but merely gives a reason for placing limits on the extent of sacrifice that any generation should be expected to make.[9] "Our belief is not that the importance of future benefits steadily declines. It is rather that no generation can be morally required to make more than certain kinds of sacrifice for the sake of future generations" (Parfit 1984, p. 484). According to Parfit's objection, then, the argument from excessive sacrifice merely draws attention to the fact that benefits and costs should be distributed among generations in an equitable manner: "we should not simply aim for the greatest net sum of benefits. We should have a second moral aim: that the benefits be fairly shared between different generations" (Parfit 1984, p. 485). Furthermore, Parfit argues, misrepresenting a moral judgment about what constitutes a fair distribution of sacrifices as a pure time preference would lead to undesirable conclusions.

> Suppose that, at the *same* cost to ourselves now, we could prevent either a minor catastrophe in the nearer future, or a major catastrophe in the further future. Since preventing the major catastrophe would involve no extra cost, the Argument from Excessive Sacrifice fails to apply. But if we take that argument to justify a Discount Rate, we shall be led to conclude that the greater catastrophe is less worth preventing. (Parfit 1984, p. 485, italics in original)

In short, limiting the extent of sacrifices that may be demanded by morality is not equivalent to a pure time preference.

I agree with Parfit that the argument from excessive sacrifice fails as an argument for the pure time preference.[10] However, a version of the argument proposed by Kenneth Arrow (1999) does raise an interesting philosophical challenge for the presumption of universalizability as a foundation for intergenerational impartiality. Arrow frames the argument from extreme sacrifice as a rejection of universalizability in favor of agent-relative ethics (1999, p. 16). An ethical theory is agent-neutral if it states that all

[9] Similar replies to the argument from excessive sacrifice are voiced in Cowen and Parfit (1992, pp. 148–9), Broome (1992, pp. 105–6), and Gardiner (2011, pp. 294–5). Broome (1992, pp. 104–5) also critiques one of the technical assumptions underlying the argument.

[10] See section 5 of the Appendix for a critical discussion of Chichilnisky's (1996) version of argument from extreme sacrifice.

agents have the same moral aims, while it is agent-relative if it gives different aims to different individuals (Parfit 1984, p. 27). Consequentialism is the classic example of an agent-neutral ethical theory, since it states that every individual is morally bound to promote the best overall consequences. Ethical egoism, according to which the moral responsibility of each person is to promote his or her own interests, is a simple example of an agent-relative theory – it says that *my* aim is to promote *my* interests, and that *your* aim is to promote *yours*. A less egoistic agent-relative ethical theory might claim that each person has duties to promote the interests of people with whom he or she bears special relationships, such as friends and family. Like ethical egoism, such a theory would assign distinct moral aims to distinct people. Arrow cites Samuel Scheffler's (1994) book, *The Rejection of Consequentialism*, as an example of the sort of agent-relative ethics that is suited to his argument. Scheffler proposes to modify consequentialism with what he calls "agent-centered prerogatives" (1994, p. 5). These prerogatives permit people to behave according to their natural inclinations in some circumstances even when doing so is not what would promote the best consequences overall, for example, by devoting resources to one's own children rather than to famine relief. Similarly, Arrow suggests, agent-centered prerogatives provide a philosophical rationale for people to value the present more heavily than the future, thereby supplying a rationale for a pure time preference.

Linking a pure time preference to an agent-relative ethical perspective draws attention to the fact that universalizability is not an entirely uncontroversial basis for moral reasoning. If universalizability is to be invoked in defense of intergenerational impartiality, therefore, some case needs to be made for why it should be a component of the concept of fairness in the context of intergenerational decisions. To pursue the agent-relative challenge to intergenerational impartiality, two additional issues should be considered. First, what sorts of pure time preference do people actually have? This question is important if one accepts Scheffler's notion of agent-centered prerogatives, wherein people are granted dispensations from the demands of consequentialism to pursue their own natural inclinations to a certain extent. It is also important if one adopts a descriptive approach to discounting. The second issue is that the argument from excessive sacrifice is not the most promising approach for one who wishes to use agent-relative ethics in support of a pure time preference. Instead, agent-relative ethics is more plausibly construed in terms of a different argument, known as the argument from special relations (Parfit 1984, pp. 485–6). These two issues are taken up in the next section.

### 6.3.2 *Actual time preferences and special relations*

It seems obvious that people often do value the present more than the future. People are generally more concerned about what will happen to them this year than about what will transpire twenty years from now and are more concerned about events that will happen in their own lives than those occurring two centuries later. Given a descriptive approach to discounting, such observations about human nature would constitute an argument for a pure time preference, that is, for $\rho > 0$ in the Ramsey equation. I consider two objections to this argument. The first is the claim that a pure time preference is irrational (Cowen and Parfit 1992, pp. 154–5). The second objection is that people's actual pure time preferences do not support exponential discounting, that is, discounting by multiplying a fixed rate at each unit of time as expressed in the equation $PV(x_t) = x_t/(1 + r)^t$. Let us consider these objections in turn.

Why is the pure time preference irrational? Parfit's argument for this claim rests on the following proposition:

> A mere difference in *when* something happens is not a difference in its quality. The fact that a pain is further in the future will not make it, when it comes, any less painful. (1984, p. 164, italics in original)

The irrationality, then, arises from the fact that a pure time preference may result in a person choosing a more severe harm in the distant future over a less severe harm in the near future. Such behavior is said to be irrational because the person knowingly and for no reason chooses an option that she knows will result in her being less well off (cf. Parfit 1984, p. 124).

But Parfit's reasoning cannot be transferred to an intergenerational context without further substantive premises. For in the intergenerational case, the deferred harms may be imposed on other people. Moreover, it seems very likely that people at present do possess a pure time preference with respect not only to their own future but also to future generations. Indeed, it is likely that they care significantly less about future generations than about future portions of their own lives. So, an additional argument would be needed to explain why a pure time preference should be regarded as irrational in the intergenerational context. One might object that a pure time preference across generations is immoral (see Parfit 1984, pp. 480–1). While I sympathize with this response, it merely amounts to asserting intergenerational impartiality, which of course is the issue in dispute.

Turn, then, to the second objection, which points out that people's actual pure time preferences are not consistent with exponential discounting. A person who exponentially discounted costs and benefits at a constant rate, say 5%, applied on an annual basis would exhibit a certain type of neutrality with regard to the passage of time. Such a person would discount the benefits and costs of each year by 5% in comparison to the preceding year. Thus, a benefit obtained this year would be discounted by 5% in comparison to a similar benefit obtained last year. Similarly, a benefit obtained 101 years from now would be discounted by 5% in comparison to one obtained 100 years hence. Exponential discounting is a straightforwardly sensible procedure with regard to private investment decisions. If I can invest money at an annual rate of return of 5%, then there is always a potential financial advantage to having a given amount of money earlier rather than later. And if I must decide between $\$x$ now and $\$y$ ten years in the future, then the latter option is more lucrative only if $x < y/(1 + 0.05)$. An exponential rate of discount is also the standard assumption in cost–benefit analysis, as noted in section 6.2.

But pure time preferences that people actually have depart from exponential discounting in two fundamental respects (Price 1993, pp. 110–13). First, people normally do not discount the future at a constant rate. Instead, they usually discount the near future at a high rate and then are relatively indifferent between times in the distant future, a pattern known as hyperbolic discounting. Second, they generally do not value the past more highly than the present. To the contrary, they often devalue the past to an even greater extent than the future. As a result, people's actual pure time preferences do not provide any argument for exponential discounting. Indeed, if one adopted a descriptive approach, actual preferences might be used to argue *against* exponential discounting (see Pearce, Atkinson, and Mourato 2006, pp. 185–6). But the attitude of economists toward hyperbolic discounting is usually not one of deference. Instead, such discounting is typically regarded as irrational because it leads to time inconsistencies. To see the idea, consider this hypothetical example:

> When offered, say, a four-week world cruise in year 10, or a five-week world cruise in year 11, they probably prefer the longer cruise, because year 10 and year 11 both fall in the uniformly perceived distant future. However, when year 10 arrives the perception has changed. Year 10 is the present, and the extra week is too far away to be worth waiting for. Yet from the perspective of year 0 this would seem irrational. (Price 1993, p. 103)

A basic argument for exponential discounting, then, is that it precludes time inconsistencies of this sort.

The pure time preferences of actual people, then, provide no grounds for exponential discounting. Nor does the existence of such preferences, considered in isolation, constitute much of an objection to intergenerational impartiality. Actual preferences need not be treated as normative, especially when there is some reason to think that those preferences are unreasonable. However, the pure time preferences of actual people, I suggest, take on a more interesting and – from the perspective of intergenerational impartiality – more challenging character when considered in connection with the argument from special relations.

The argument from special relations is founded on a commonsensical idea that Parfit states as follows:

> . . . there are certain people to whose interests we *ought* to give some kinds of priority. These are the people to whom we stand in certain special relations. Thus each person ought to give some kinds of priority to the interests of his children, parents, pupils, patients, those whom he represents, or his fellow citizens. Such a view naturally applies to the effects of our acts on future generations. Our immediate successors will be our own children. According to common sense, we ought to give their welfare special weight. We may think the same, though to a reduced degree, about our children's children. (1984, p. 485, italics in original)

The argument from special relations, therefore, proposes a moral basis for favoring the interests of closely following generations over those in the more distant future. In so doing, it suggests a rationale for a pure time preference regarding future generations: it is reasonable to care more about nearer times because we have special relationships with people living then that we do not have with people living at later times.

This reasoning is also related to the agent-relative ethics challenge to universalizability. That is, it proposes that each generation has a distinct moral aim, namely to promote the interests of *its* successors and, to a lesser degree, its successors' successors. Wilfred Beckerman and Cameron Hepburn (2007) also discuss the connection between the argument from special relations and agent-relative ethics:

> Thus the reasons for giving serious consideration to agent-relative ethics include (i) a long philosophical tradition stretching back at least to Hume; (ii) probably universally held public preferences; and (iii) within limits, its instrumental value. . . . The point is that, whatever the "right" answer, climate policy cannot properly be conducted without considering a range of ethical perspectives, including those that attach a lower value to a unit of welfare accruing to a distant generation as to one accruing today. (Beckerman and Hepburn 2007, p. 201)

Note that the three items highlighted by Beckerman and Hepburn track the three arguments for the pure time preference discussed here: (i) corresponds to the argument from special relations, (ii) to the argument from actual time preferences, and (iii) to the argument from excessive sacrifice. Rather than three independent arguments, then, they can be viewed as an interconnected challenge to intergenerational impartiality.

Some critics of the pure time preference regard the argument from special relations as the most serious challenge confronting intergenerational impartiality (Broome 1992, p. 108; Stern 2008, p. 15). Even Parfit acknowledges that the argument from special relations "might support a new kind of Discount Rate" (1984, p. 485). What, then, might be said in reply to this argument? Parfit suggests two responses.[11] First, he points out that the argument from special relations cannot justify exponential discounting:

> . . . more remote effects always count for less as any [exponential] Discount Rate approaches zero. But a Discount Rate with respect to Kinship should at some point cease to apply – or, to avoid discontinuity, asymptotically approach some horizontal level that is above zero. (1984, pp. 485–6)

That is because, while obligations to kin may be prioritized to some extent, it is not morally acceptable to regard others as having no significance whatever. There is some level of moral regard that should be shown to all people, whether one stands in special relation with them or not. However, this response, while correctly pointing out that the argument from special relations does not support exponential discounting, also surrenders intergenerational impartiality. Indeed, some climate change economists have suggested a discount rate that, like Parfit's "new kind of Discount Rate," declines with time in part because it would more faithfully correspond to people's actual hyperbolic time preferences (see Pearce *et al.* 2006, p. 23). Yet, as Gardiner points out, whether and to what extent declining discount rates reduce intergenerational injustices resulting from discounting depends on the numerical values chosen (2011, pp. 287–8). Parfit's first reply, therefore, is not adequate for one who wishes to defend intergenerational impartiality.

Parfit's second objection to the argument from special relations is that it "does not apply to the infliction of grave harms" (1984, p. 486). Parfit supports this claim with the following example:

> Suppose [the US] Government decides to resume atmospheric nuclear tests. If it predicts that the resulting fall-out would cause several deaths, should

[11] These replies are reiterated in Broome (1992, pp. 107–8) and Cowen and Parfit (1992, pp. 149–50).

> it discount the deaths of aliens? Should it therefore move the tests to the Indian Ocean? I believe that, in such a case, the special relations make no moral difference. We should take the same view about the harms that we impose on our remote successors. (1984, p. 486)

This is an intriguing objection and one that might function as an argument against applying a pure time preference to potential catastrophes in the distant future. In this vein, Broome suggests that Parfit's reasoning is sufficient to answer the argument from special relations for the purpose of climate change policy. However, Parfit's reply suffers from two difficulties. First, the distinction between grave and non-grave harms is very vague. That vagueness is crucially relevant in the case of climate change, wherein adverse effects are expected to become gradually more severe over time. Since under the present proposal discounting would be allowed up until the point at which the harms become "grave," where one draws that line could have significant implications for decisions about climate change mitigation. The second difficulty is that, just as his first, Parfit's second reply to the argument from special relations abandons intergenerational impartiality. Indeed, from this perspective the two replies are not all that different. Each grants that a pure time preference across generations may be justifiable to some degree but insists there are limits, left rather vague, on what such a time preference could justify. Although this response rejects the standard approach of a pure time preference combined with exponential discounting, it is compatible with approaches to discounting that use a declining rate. Thus, the challenge to intergenerational impartiality raised by the argument from special relations remains unanswered.

## 6.4 A new argument for intergenerational impartiality

This section develops a defense of intergenerational impartiality based upon the concept of a sequential plan. Section 6.4.1 elaborates the notions of sequential plan and sequential justifiability, while section 6.4.2 explains how these lead to an argument for intergenerational impartiality that is capable of answering the challenge posed by agent-relative ethics. A distinctive feature of this argument is that it defends intergenerational impartiality on the basis of practical considerations pertinent to effectively implementing sequential plans rather than by appeal to fundamental principles of ethical theory. It argues that sequential plans should not discriminate against future generations because this makes them unlikely to be accepted by future actors upon whom successful completion of the plan depends.

### 6.4.1 *Sequential plans and sequential justifiability*

Many, if not most, environmental policies require a long-term commitment to achieving a goal, as well as a commitment to a structure or mechanism through which that goal is to be attained. For example, consider the Montreal Protocol on Substances that Deplete the Ozone Layer, which arose in response to scientific evidence that human emissions of halogenated hydrocarbons were depleting stratospheric ozone. This international agreement was signed in 1987, came into effect in 1989, and underwent seven revisions during the 1990s. At present, it is expected that this agreement, if followed through, will restore the stratospheric ozone layer to its pre-1980 state sometime between 2050 and 2060.[12] Of course, the Protocol will continue to be relevant after that date, as it would be unwise to revert to the widespread use of ozone-depleting substances.

I use the term *sequential plan* to refer to a plan consisting of a series of steps enacted over an extended period time. The Montreal Protocol is an excellent example of a sequential plan. It is a plan – that is, it is directed at particular aim (i.e., the restoration of stratospheric ozone) promoted via a specific method (i.e., phasing out a list of ozone-depleting substances) – and its implementation must occur over a long expanse of time. In the case of the Montreal Protocol, this time frame is, at a minimum, sixty to seventy years. However, one could plausibly argue that the future extent of the Protocol is open-ended because maintaining restrictions on ozone-depleting substances will remain important even after the stratospheric ozone layer is restored. Given the extensive timescale of the plan, it is obvious that it cannot be implemented all at once. Like the Montreal Protocol, any serious approach to climate change mitigation, whether at the national or international level, would also take the form of a sequential plan, a point illustrated by the example of a carbon tax "ramp." Similarly, any environmental policy that phases out a problematic substance, such as DDT or lead solder in electronics, is in effect a sequential plan. Even if the initial phase-out is relatively quick, the plan implies a commitment to maintaining restrictions on the use of the substance for the foreseeable future.

Well-designed sequential plans are often adaptive in the sense discussed in section 3.4.1. But my concern in this section is with another feature of sequential plans that is even more fundamental to their success, namely sequential justifiability. This concept arises from the simple fact that those

[12] See the World Meteorological Organization Global Ozone Research and Monitoring Project – Report No. 52, *Scientific Assessment of Ozone Depletion: 2010*.

living in the present do not have the power to control the actions of future people. In particular, those who implement the first steps of a sequential plan cannot compel future actors to follow through on the remainder. Thus, a sequential plan will achieve its aim only if future actors continue to be committed to following its directives, for example, by maintaining a commitment to implementing the scheduled phasing out of ozone-depleting substances.[13] Yet persistent commitment to following through on a sequential plan may be difficult to secure for a number of reasons, one of which is that the preferences and priorities of people are likely to change over time. An environmental challenge that rose to the forefront in the 1980s might no longer be a front-burner concern in 2030.

A plan is *sequentially justifiable* if and only if it can be justified to actors implementing the plan at every stage. Since preferences can be expected to change across generations, it is unwise to base a sequential plan upon preferences that are likely to be rejected by future people who will be responsible for implementing the plan. Instead, a plan is sequentially justifiable when good reasons can be given at every step for continuing to carry out the plan. To clarify this idea, I define two types of reasons: agent-relative and burden shifting.

I will say that *a reason* R *is agent-relative with respect to a person or group* A just in case R is a reason for A to act in a particular way but not necessarily a reason for others to do so. To take a simple example, the fact that mango-coconut is my favorite flavor of ice cream is a reason for me to choose it but not necessarily a reason for others. The definition says "not necessarily" instead of "not" because it is possible that an agent-relative reason for me might be a reason for others, too. For instance, suppose that others wish to mimic my tastes (imagine that I am a widely admired ice cream connoisseur whom some wish to emulate in order to appear culturally sophisticated). The significance of an agent-relative reason is that it is *optional* for others whether it is a reason for them. In other words, if R is agent-relative with respect to A, but B rejects R, then there is no reasoned argument that could be provided to persuade B to be motivated by R.

Consider this idea in relation to the pure time preference. Suppose that R states that the sequential plan is optimal given A's actual preferences, which include a pure time preference that substantially devalues the interests of future generations. Then R is plausibly a reason for A to accept that sequential plan. But future people almost certainly will not share A's

[13] This in fact continues to be a serious challenge in the case of the Montreal Protocol (see Gareau 2013).

pure time preference, that is, they will not devalue their own present in comparison to A's. And since A's reason is agent-relative, there is no justification that could be given to them for adopting such a preference. While the most likely situation is that a sequential plan would be designed in accordance with reasons that are agent-relative with respect to those who initiate it, reasons might be agent-relative with respect to actors at any stage of a sequential plan. Recall from the previous section that people often discount the past to an even greater extent than they discount the future. Thus, actors at the end stages of the plan might prefer that the costs of implementation would have been overwhelmingly borne by earlier actors. Yet such preferences, while perhaps agent-relative reasons with respect to those future people, would not be shared by earlier actors and would not constitute reasons for them.

However, it would be a mistake to suggest that agent-relative reasons always have an undermining effect on sequential plans. Suppose that members of the initiating generation feel a duty to remediate an environmental problem they have caused and therefore take greater costs upon themselves than they demand of future generations. This duty would be an agent-relative reason with respect to the initiating generation: it is a reason for them to take on additional burdens but not a reason for other generations. But it is unlikely that later generations would deem the plan unacceptable if the initial generation had decided to act upon this duty. I suggest, then, that the problem is not agent-relative reasons per se but agent-relative reasons that are also burden shifting. I will say that *a reason is burden shifting from a person or group* A *to another* B if it is a reason for shifting costs away from A and onto B. Notice again that burden-shifting reasons are not necessarily problematic, since there may be legitimate reasons why some should shoulder more costs than others. The problem arises when a sequential plan is motivated by *discriminatory* reasons that are both agent-relative and burden shifting.

I will say that R *is a discriminatory reason against a person or group* B if and only if R is agent-relative with respect to some person or group A distinct from B and is burden shifting from A to B. Discriminatory reasons are important because they undermine sequential justifiability. When a sequential plan is driven by discriminatory reasons, the preference of generation A to shift burdens to generation B is unlikely to be shared by B, and since the reasons underlying that preference are agent-relative with respect to A, no justification can be provided to B for accepting them.

I believe that sequential justifiability is an absolutely crucial feature of sequential plans because, without it, there is no basis for expecting that a sequential plan will be effectively implemented. In the next section, I explain how sequential justifiability leads to an argument for intergenerational impartiality.

### 6.4.2 *Discrimination and undermining sequential plans*

In this section, I develop an argument for intergenerational impartiality on the basis of the concepts of sequential plan, sequential justifiability, and discriminatory reasons. The core of this argument rests upon two premises. The first is that the pure time preference is a discriminatory reason in the context of a sequential plan. The second is that a sequential plan is more difficult to effectively implement when it is based upon reasons that discriminate against future actors responsible for bringing it to completion. Since intergenerational impartiality is equivalent to the absence of a pure time preference, these two premises lead to the following conclusion:

> (SP) *If* you are committed to implementing an effective sequential plan, *then* you should presume intergenerational impartiality in the design of that plan.

That is, the two premises entail that a pure time preference is an obstacle to the effective implementation of sequential plans. Thus, it follows that if one's goal is to design a sequential plan that will be effectively implemented, pure time preferences should be avoided, which is just to say that intergenerational impartiality should be assumed instead. The implications of (SP) are examined below. For now, consider the rationale of the two premises.

The first premise is a straightforward consequence of the definition of a discriminatory reason: a pure time preference is an agent-relative reason of the present for shifting burdens to future generations. The rationale for the second premise stems from the fact that present people do not have the power to compel actors responsible for implementing future stages of a sequential plan. Thus, the effective implementation of a sequential plan depends on its acceptance by those actors. But these future people are unlikely to accept a plan designed in accordance with reasons that discriminate against them. Future generations will not share the present's pure time preference – to the contrary, we can expect that they will be inclined to regard their own present as more valuable than the past. And since the pure time preference is an agent-relative reason, no justification can be given to them for adopting the pure time preference of a previous generation.

Consider this point in relation to a hypothetical climate change mitigation policy that requires little action on the part of the present while deferring costly measures to future generations. At the first step of the plan some minimal and inexpensive steps are taken (e.g., providing some funding for research on alternative energy), while the plan calls for subsequent generations to drastically reduce their GHG emissions. Moreover, part of the rationale for this plan is a pure time preference of the present generation, which is used to support the contention that later generations should shoulder a greater proportion of the costs of climate change mitigation. But no rational argument could be given to these future actors for accepting reasons that discriminate against them in this way. No good reason could be given to them for valuing the past more highly than their own present. After all, it is entirely arbitrary which moment in time is selected as the undiscounted vantage point from which a pure time preference is applied to all subsequent times. There is no reason, then, for future actors to agree that this privilege should be bestowed upon some earlier time rather than on their own present. Consequently, there is no reason for decision makers at later stages of the sequential plan to incur upon themselves the disproportionate costs required by a plan founded upon the application of a pure time preference at an earlier date. Indeed, this behavior of their predecessors is most naturally taken as an encouragement for them to follow suit and similarly "pass the buck" along to the next generation.[14]

Moreover, declining discount rates do not substantively alter this argument. A common objection to exponential discounting is that it entails that extreme catastrophes in the distant future count for practically nothing at present, a result that Gardiner refers to as the "absurdity" problem (Gardiner 2011, p. 268). Declining discount rates are a means of avoiding this absurdity. A discount rate that declines and then goes to zero at some point in the future (see Weitzman 2001, p. 270) will place limits on the extent to which future generations can be discounted from the perspective of the present. However, even if applied at a declining rate that converges to zero, a pure time preference is an agent-relative reason for shifting burdens from the present to the future and is, therefore, discriminatory. As a result, it is a highly undesirable feature of a sequential plan for the reasons explained above. Moreover, there is a sense in which declining discount rates are even *more* discriminatory than exponential discounting. Exponential discounting is fair in one peculiar sense: it permits each generation

[14] See Gardiner's discussion of "intergenerational buck passing" (2011, chapter 5).

to discriminate against the next at a constant rate. In contrast, declining discount rates would permit discounting by a group of generations close to the present but would forbid all later generations from doing so. But what reason could be given to later generations to accept a hypocritical plan that discriminates against them while sternly prohibiting them from doing the same to their own successors?

One might object to the argument for (SP) that a pure time preference can facilitate the implementation of sequential plans if the most difficult step is the first one. Perhaps once the initial hurdles to taking action have been overcome and institutions have been put in place, it will be much easier to move forward, for instance, by incrementally increasing the rate of a carbon tax. The pure time preference, then, might be defended on the grounds it makes that first step easier to take. However, this argument fails to support a pure time preference. Instead, it is an argument that it may often be wise to design sequential plans so that they gradually introduce more demanding requirements, as in the case of a carbon tax ramp. There are several reasons for such an approach, none of which bear any relation to devaluing the future simply because it is the future. One reason is that the effects of the plan may be difficult to predict. In such cases, it is reasonable to proceed initially in small steps before proceeding to more aggressive action. A second reason is that the first initial steps may create incentives for social or technological changes that will make larger steps easier later on. Both of these reasons are relevant to a carbon tax. Economic effects of a carbon tax may be difficult to anticipate, and the presence of the tax provides an incentive for behaviors and technologies that reduce reliance on carbon-based fuels.

If concerns such as these are the real motivation for a carbon tax ramp, then those reasons – and not a spurious pure time preference – should be explicitly given as its justification (see Cowen and Parfit 1992). Unlike a pure time preference, the two reasons just sketched for gradually phasing in a sequential plan are fully compatible with intergenerational impartiality. Both point to relevant differences between the present and future: greater knowledge about the effects of the policy approach in the first case and technological or behavioral innovations in the second. Thus, these reasons do not undermine the sequential justifiability of the plan. Moreover, explicitly stating the actual justification for the sequential plan allows it to be subjected to careful examination. That matters because the importance of substantive reasons for gradually implementing a sequential plan can vary from case to case, and hence those reasons would not be accurately reflected by a uniformly applied pure time preference.

Let us turn, then, to the implications of (SP). Advocates of PP clearly would accept the commitment to sequential plans in the antecedent of (SP), so by modus ponens they should accept intergenerational impartiality, too. This should come as no surprise, since disregard for the future obviously runs counter to precaution.[15] But (SP) is also relevant to people who are not necessarily fans of PP. For example, consider an economist who devotes a significant portion of her career to attempting to estimate the optimal carbon tax ramp. It would be difficult to understand why a person would dedicate so much of her life to such a project if she were not committed to designing sequential plans that would be effectively implemented. So (SP) draws out a deep incoherence in economic analyses of optimal policies for addressing climate change, or any other long-term problem, that assume a non-negligible pure time preference.

However, there can be little doubt that not everyone shares the commitment to sequential plans specified in the antecedent of (SP). People who care little for future generations would also be unconcerned about whether those generations follow through on sequential plans aimed at mitigating long-term environmental problems. But there is a simple ethical argument against such an attitude. That ethical argument arises from the response Parfit makes to the argument from special relations, namely that there are limits to what appeals to special relations can ethically justify. We may be ethically justified in favoring those with whom we share special relations, but it is not morally permissible to completely disregard the effects of our actions upon everyone else. In the context of long-term environmental problems, such as climate change or stratospheric ozone depletion, this creates an obligation to implement effective sequential plans.

Given (SP), therefore, an ethical argument for intergenerational impartiality rests on the relatively minimal ethical claim that the present has a moral obligation to develop sequential plans to address long-term environmental and health issues. This argument does not assume universalizability as a premise and is compatible with agent-relative ethics. Since the argument is restricted to sequential plans in the realm of environmental and health policy, it does not require intergenerational impartiality in other contexts, such as personal financial decisions. In addition, even within its intended context, the argument does not prohibit agent-relative reasons generally but only discriminatory reasons. Consequently, this defense of

[15] Hartzell (2009, p. 154) and Kysar (2010, p. 257) also associate PP with a rejection of the pure time preference.

intergenerational impartiality defuses the agent-relative challenge raised in section 6.3.

In addition, (SP) has implications for the contrast between descriptive and prescriptive approaches outlined in section 6.1. A descriptive approach is committed to assessing sequential plans on the basis of actual preferences. Since future people do not yet exist, in practice that means the preferences of present people, as best as these can be estimated. Even putting aside the many conceptual and practical conundrums surrounding inferring preferences from behavior (see Sagoff 2004, 2008), the above discussion highlights a serious difficulty for the descriptivist approach in connection with sequential plans. For the reasons given above, basing a sequential plan on the actual preferences of present people is problematic from an instrumental perspective – if the aim is genuinely to craft an effective plan.

## 6.5 Conclusions

In this chapter, I have suggested that the standard contrast between descriptive and prescriptive approaches to discounting camouflages a powerful practical argument for intergenerational impartiality. Long-term environmental problems usually must be addressed by sequential plans, that is, plans enacted in stages over an extended period of time. And intergenerational impartiality is an important element of such sequential plans from an instrumental perspective. A sequential plan is unlikely to be effectively implemented when it is designed on the basis of reasons that discriminate against actors who will be relied upon to bring it to completion. One advantage of this argument is that it avoids reliance on universalizability and hence cuts off the challenge to intergenerational impartiality raised by agent-relative ethics. The result of this chapter, then, is analogous to that of Chapter 4, which argued that a central argument for PP could be made in a manner that is independent of controversial propositions concerning meta-ethics.

CHAPTER 7

# *Precautionary science and the value-free ideal*

## 7.1 Science and values

According to some advocates of PP, a thoroughly precautionary approach to environmental policy would require a precautionary science to go along with it (Barrett and Raffensperger 1999; Kriebel *et al.* 2001; Tickner and Kriebel 2006). Some proposals made in the name of precautionary science, although perhaps difficult to implement in practice, seem relatively uncontroversial in principle. For instance, advocates of precautionary science typically call for a greater emphasis on interdisciplinary science, greater transparency, and an increased role for public participation in shaping research agendas of policy relevant science (Barrett and Raffensperger 1999, p. 109; Kriebel *et al.* 2001, p. 875). In this chapter, I examine the more controversial idea that value judgments relating to human or environmental health can legitimately influence scientific inferences – for instance, about the toxicity of a chemical. More specifically, advocates of PP often insist that value judgments can – and indeed inevitably *must* – influence decisions about what is to count as sufficient evidence for accepting a claim (Barrett and Raffensperger 1999; Kriebel *et al.* 2001, pp. 873–4; Sandin *et al.* 2002, p. 295). In contrast, several critics of PP argue that such an influence of non-epistemic values would constitute an unacceptable bias in the interpretation of scientific evidence (Clarke 2005; Harris and Holm 2002).

In philosophy of science, the position defended by these critics is known as the *value-free ideal*, according to which "social, ethical, and political values should have no influence over the reasoning of scientists, and that scientists should proceed in their work with as little concern as possible for such values" (Douglas 2009, p. 1). In this chapter, I discuss what I consider to be the most significant objection to the value-free ideal, namely the argument from inductive risk, according to which non-epistemic values *should* influence standards of evidence required for accepting or rejecting hypotheses when there are significant social costs associated with errors.

This argument was a lively topic of dispute in the philosophy of science literature in the 1950s and 1960s (Churchman 1956; Hempel 1965; Jeffrey 1956; Levi 1960, 1962, 1967; Rudner 1953), fell from interest for about two decades, and was revived in the 1990s (Biddle and Winsberg 2010; Cranor 1993, 2006; Douglas 2000, 2009; Elliott 2011a; Kitcher 2001, 2011; Kourany 2010; Shrader-Frechette 1991; Steel 2010, 2013b; Steel and Whyte 2012). Moreover, some defenders of PP explicitly cite the argument from inductive risk (Sandin *et al.* 2002, p. 294). However, the argument from inductive risk remains controversial in two general respects. First, the argument remains subject to a number of challenges and objections. Second, since not all manifestations of values in science are salutary, if the argument is accepted, one is confronted with the challenge of articulating a new normative standard to replace the value-free ideal.

Section 7.2 presents the argument from inductive risk, while section 7.3 examines and replies to objections to it. Probably the most common objection is that the argument from inductive risk presupposes a behavioral notion of acceptance that is inappropriate for science (Dorato 2004; Giere 2003; Lacey 1999; Levi 1967; McMullin 1982; Mitchell 2004). A concept of acceptance is behavioral if it entails that one can accept a statement only in relation to some practical objective, such as reducing the incidence of cancer. In section 7.3.1, I answer this objection by showing that the theory of acceptance developed by Jonathan Cohen (1992) supports the argument from inductive risk without interpreting acceptance in a behavioral manner. According to Cohen, to accept a proposition in a particular context is to make it available as a premise for reasoning in that context. Cohen's concept of acceptance is cognitive rather than behavioral, but it nevertheless supports the claim that ethical consequences of error can legitimately influence decisions about what constitutes sufficient evidence. In section 7.3.2, I argue moreover that Cohen's theory is preferable to alternative theories of acceptance according to which acceptance should be guided solely by epistemic values. These alternatives can be divided into two groups. The first consists of proposals that associate rational acceptance with maximizing expected cognitive utility (Harsanyi 1983; Levi 1967; Maher 1993). The second group of alternative theories insists that certainty is a necessary condition for rational acceptance (Lacey 1999; Levi 1980). Theories in the first group claim that what we accept should have no influence on what we do. I argue that this position is highly problematic when one considers the possibility of accepting normative as well as descriptive propositions, which can then jointly function as premises in practical or ethical reasoning. Theories in the second group have the significant shortcoming of linking acceptance to dogmatism. By contrast, Cohen's theory of acceptance suffers from neither of these difficulties. Finally, section 7.3.3 critically examines the objection, due to Richard Jeffrey (1956), that scientists should not accept hypotheses at all but instead merely make probability judgments concerning hypotheses that they pass along to policy makers.

Rejecting the value-free ideal requires a reevaluation of the proper role of values in science and of the sorts of influences that threaten scientific integrity. The proposal I defend on this topic relies on a distinction between epistemic and non-epistemic values. That distinction is very contentious in philosophy of science in part because it is often associated with the value-free ideal (Douglas 2007, 2009; Longino 1996; Rooney 1992). As Douglas puts it, "A clear demarcation between epistemic (acceptable) and non-epistemic (unacceptable) values is crucial for the value-free ideal" (2009, pp. 89–90). And indeed, proponents of the ideal do invoke the distinction in defense of their views (Lacey 1999; McMullin 1982). It is tempting, therefore, to assume that a critic of the value-free ideal must also object to the distinction between epistemic and non-epistemic values. In section 7.4, I argue that this assumption is not only mistaken but also unfortunate because it obscures the usefulness of the epistemic/non-epistemic distinction for a positive account of the role of values in science. I develop the idea that epistemic values are distinguished on the grounds that they promote the attainment of truth and defend this proposal against a number of common objections. The distinction between epistemic and non-epistemic values is the basis for a replacement norm for the ideal of value-free science proposed in Chapter 8.

## 7.2 The argument from inductive risk

The argument from inductive risk challenges the idea that ethical values should have no influence on what inferences scientists make from data, an idea that Hugh Lacey calls *impartiality* (1999, pp. 69–70). Impartiality states that the acceptance or rejection of scientific hypotheses should, ideally, be guided by epistemic values only and should not be influenced by non-epistemic concerns. The distinction between epistemic and non-epistemic values is the topic of section 7.4. For now, suffice to say that epistemic values are supposed to be truth promoting, while non-epistemic values need not be. The preference for scientific theories that make accurate predictions would be an example of an epistemic value, while promoting public health would be an ethical rather than epistemic value. Impartiality is distinct from *autonomy* and *neutrality*, two other ways in which science might be thought to be value free (Lacey 1999, pp. 74–87). Autonomy concerns the extent to which the research problems and questions of science should be driven by the intellectual curiosity of scientists rather than by broader social imperatives, while neutrality claims that science cannot provide answers to moral or political questions. The distinctions among impartiality, autonomy, and neutrality are important because reasons for rejecting autonomy and neutrality are not necessarily arguments against impartiality. Thus, that scientists' choices of research topics are often influenced by available funding sources, which in turn are influenced by non-epistemic values of the funders, is an objection to autonomy, not impartiality. Similarly, that scientific research (e.g., about the relationship between legalized abortion and crime rates) sometimes has unavoidable implications for controversial political and moral issues is an argument against neutrality, not impartiality. Since Max Weber (1949), a common strategy of defenders of the value-free ideal is to grant that autonomy may fail but to nevertheless defend impartiality (Lacey 1999). The argument from inductive risk, then, merits attention because it directly challenges this last line of defense of the ideal of value-free science.

The classic statement of the argument from inductive risk in the philosophy of science literature occurs in a (1953) paper by Richard Rudner titled, "The Scientist Qua Scientist Makes Value Judgments."[1] After explaining why he finds previous arguments on both sides of the issue unsatisfactory, Rudner proceeds to offer what he describes as "a much stronger case ... for the contention that value judgments are essentially involved in the procedures of science" (1953, p. 2). That argument leads off with the following assertion:

> Now I take it that no analysis of what constitutes the method of science would be satisfactory unless it comprised some assertion that the scientist as scientist accepts or rejects hypotheses. (1953, p. 2)

Rudner then discusses several examples to illustrate that decisions about what should count as sufficient evidence for accepting a hypothesis depends on the practical or ethical consequences of being mistaken. For instance, he suggests that a level of certainty that suffices for accepting that a belt-buckle stamping machine is not defective would not be enough to justify accepting that the level of a toxic constituent in a drug is sub-lethal (1953, p. 2). These examples lead to the second premise of Rudner's argument:

[1] Several other philosophers also published versions of the argument from inductive risk in the same period (see Braithwaite 1953, chapter 7; Churchman 1948; Hempel 1965; Nagel 1961).

> In general then, before we can accept any hypothesis, the value decision must be made in the light of the seriousness of the mistake, that the probability is *high* enough or that, the evidence is *strong* enough, to warrant its acceptance. (1953, p. 3)

From these two premises, Rudner concludes that value judgments about the seriousness of mistakes are an unavoidable aspect of accepting a scientific hypothesis. Rudner notes that this reasoning is implicit in widely used statistical significance tests (1953, p. 3), wherein one decides which of the two errors is worse (the Type I error), sets the rate of that error at an acceptably low level, and then makes probability of the other error (Type II) as low as possible given the error rate of the first. Normally, the Type I error consists of rejecting the null hypothesis when it is in fact true. For instance, in assessing the toxicity of a chemical, the null hypothesis might be that a specific level of exposure to the chemical has no toxic effect. The Type II error would then be the failure to reject the null hypothesis when it is in fact false. References to Type I and Type II errors remain common in statements of the argument from inductive risk, including those found in the literature on PP (Kriebel *et al.* 2001, pp. 873–4; Peterson 2007, pp. 7–8; Sandin *et al.* 2002, pp. 294–5).

It will be helpful to have a concise paraphrase of the argument from inductive risk in the language used here:

1. One central aim of scientific inference is to decide whether to accept or reject hypotheses.
2. Decisions about whether to accept or reject a scientific hypothesis can have implications for practical action, and when this happens, acceptance decisions should depend in part on non-epistemic value judgments about the costs of error.
3. Therefore, non-epistemic values can legitimately influence scientific inference.

For instance, we might think that it is worse for ethical reasons to mistakenly reject the hypothesis that an artificial sweetener, such as saccharin, is carcinogenic than to make the opposite mistake. In that case, we should demand a higher standard of evidence for rejecting the hypothesis than for accepting it. Heather Douglas (2000) further develops the argument from inductive risk by pointing out that acceptance decisions are not merely involved in scientific inferences at the tail end of the process but are also embedded in judgments concerning methodology and modeling that give rise to the data in the first place. Thus, the argument from inductive risk should not be taken to suggest that values enter into scientific inferences only after the data are compiled and one asks whether they are sufficient to establish a particular conclusion. If the argument is sound, it entails that values may be unavoidably integrated into scientific inferences from start to finish.

## 7.3 Objections and replies

This section responds to objections challenging premises (1) and (2) of the argument from inductive risk. Sections 7.3.1 and 7.3.2 consider the objection that premise (2) depends on a behavioral concept of acceptance rather than a supposed purely cognitive understanding of acceptance claimed to be appropriate for science. The first part of the response, given in section 7.3.1, consists in describing a theory of acceptance, developed by Cohen (1992), that supports the argument from inductive risk but which is not behavioral. The second part of the response, given in section 7.3.2, is a critical examination of theories of acceptance that attempt to explain how acceptance can be guided solely by epistemic values. I argue that *any* such theory of acceptance confronts a choice among three unpalatable options: (a) claim that scientists should be relieved of the normal human responsibility to consider the consequences of one's actions; (b) deny that acceptance can (or should) have any implications for practical action; or (c) make certainty a prerequisite for acceptance. While option (a) has been critiqued by others (Douglas 2009; Kitcher 2001, 2011; Kourany 2010), I explore difficulties inherent in approaches that opt for (b) or (c). Option (b) is problematic because it is possible to accept normative as well as descriptive claims that jointly entail further claims about what ought to be done. On the other hand, option (c) implausibly links acceptance to dogmatism. Thus, the original objection can be transformed into a challenge for the objectors: provide a plausible theory according to which non-epistemic values should *not* influence decisions about accepting or rejecting scientific hypotheses. Finally, section 7.3.3 examines an objection to premise (1), which states that scientists should not accept or reject hypotheses but only form probability judgments that may be communicated to policy makers.

### *7.3.1 Accepting as premising*

The argument from inductive risk was originally inspired by Neyman–Pearson statistical theory, which allowed the practical costs of errors to influence how rates of Type I and Type II errors are set in a statistical test of a hypothesis (Churchman 1948; Rudner 1953, p. 3). This historical origin of the argument is closely related to the first objection I consider. In its classic form, Neyman–Pearson theory was associated with a behaviorist interpretation of acceptance. According to this proposal, to accept a hypothesis is simply to take some pre-specified action that would be appropriate if that hypothesis were true. One of the approach's co-founders, Jerzy Neyman, held that instead of rules of inductive *inference*, one should more properly speak of rules of inductive *behavior* (Neyman 1950). A rule of inductive behavior uses statistical data to select among a group of predefined actions, including one that corresponds to "accepting the hypothesis." For instance, if the test aims to decide whether a batch of milk is uncontaminated by harmful bacteria, accepting the hypothesis might be identified with shipping the milk off to grocery stores while rejecting the hypothesis might mean disposing of it. Arising during the heyday of behaviorist psychology and strict anti-metaphysical empiricist views in philosophy of science, Neyman's interpretation of statistical tests was very influential in its time. However, contemporary philosophical defenders of Neyman–Pearson statistics generally reject Neyman's "inductive behavior" interpretation of statistical tests (Mayo 1996). The historical origin of the argument from inductive risk, then, suggests an opening to object that it rests on an outmoded behaviorist conception of acceptance.

A commonly made objection to the argument, therefore, rejects premise (2) on the grounds that non-epistemic values should not influence acceptance construed in a cognitive, non-behavioral sense that is appropriate to science (Dorato 2004; Lacey 1999, 2004; Levi 1960, 1962, 1967; McMullin 1982; Mitchell 2004). According to this objection, premise (2) can be upheld only if acceptance is interpreted behaviorally, that is, if it is assumed that decisions about acceptance can only be made in connection with some practical decision (see Levi 1962, p. 50).[2] But, the objection continues, such an interpretation of acceptance is inappropriate for research science wherein the aim is to discover the truth rather than to achieve some practical objective.

The most direct way to answer this objection is to present a theory of acceptance that is not behavioral (i.e., which does not say that a statement can be accepted only in relation to a practical objective) but which nevertheless supports the argument from inductive risk. I suggest that a theory of acceptance proposed by Jonathan Cohen (1992) fits this bill.

[2] In fact, Sandin (2006, p. 179) insists upon a behavioral interpretation of the argument from inductive risk in the course of defending an epistemic construal of PP.

Cohen's theory centers on a distinction between acceptance and belief. According to Cohen:

> Belief that $p$ is a disposition, when one is attending to issues raised, or items referred to, by the proposition that $p$, normally to feel it true that $p$ and false that *not-p*, whether or not one is willing to act, speak, or reason accordingly. But to accept the proposition or rule of inference that $p$ is to treat it as a given that $p$. More precisely, to accept that $p$ is to have or adopt a policy of deeming, positing, or postulating that $p$ – i.e. of including that proposition or rule among one's premises for deciding what to do or think in a particular context, whether or not one feels that it is true that $p$. (1992, p. 4)

The contrast drawn here, then, is between belief as a *disposition* to feel that $p$ is true and acceptance as a *decision* to treat $p$ as a premise in a particular context. According to this perspective, acceptance and belief often coincide. For instance, I see snow falling, which causes me to believe that it is snowing, and this belief is in turn a good reason to accept that it is snowing (e.g., to use this statement as a premise in practical reasoning about how to dress).

But acceptance and belief can also come apart. Since belief on Cohen's account is an involuntary feeling, it is possible for a person to strongly feel that $p$ is true in circumstances in which she recognizes that belief in $p$ is irrational. For example, imagine a person who suffers from a dog phobia and upon seeing a friendly Boston terrier has a strong feeling that she will be viciously bitten. But she is aware that this belief is entirely unreasonable. In such cases, then, she makes a policy of treating the claim that the dog is harmless as a premise and thus makes every effort to suppress behaviors that would express the fear and alarm she so acutely feels. In this situation, Cohen's theory would say that the person believes that the dog will bite but accepts that it is harmless. Conversely, it is possible to accept something that one does not believe. For example, one might choose to take a generalization, such as the ideal gas law, as a premise in reasoning in a particular context despite believing that it is an approximation that is strictly false (Cohen 1992, p. 90).

Consider how Cohen's theory of acceptance relates to the argument from inductive risk. Cohen argues that acceptance rather than belief is the appropriate focus of scientific reasoning (Cohen 1992, pp. 86–100), which amounts to a defense of premise (1) of the argument. More significantly, Cohen's theory supports premise (2), according to which decisions about whether to accept a hypothesis can depend on ethical or practical consequences of error (see Cohen 1992, p. 12). Although Cohen does not elaborate this point in much detail, the underlying reasoning is not difficult to work out. First, note that, according to Cohen's theory, accepting a claim *p* is a decision (i.e., to adopt a policy of treating *p* as an available premise for reasoning in the context in question) that can have consequences for ethical and practical matters. The potential relevance of ethical and practical concerns to decisions about what to accept, therefore, ensues immediately from the following principle:

> *Consequence principle*: One should take the consequences of one's decisions into account in the process of deliberation.

The consequence principle does not say that consequences are the only thing that matters, nor does it specify how those consequences should be balanced or analyzed. It merely says that consequences of our acts are relevant considerations in deciding what to do. Premise (2) of the argument of inductive risk is concerned with a particular sort of consequence, namely the consequences of mistakenly accepting a claim. So to support premise (2), Cohen's theory also needs to allow that a person can accept a claim whose truth is less than certain. This is entirely possible given Cohen's theory – for instance, I may decide to accept a weather forecast as a premise in reasoning about whether to carry an umbrella while being well aware of the fallibility of meteorology. Generalizing from the specifics of Cohen's theory, then, the consequence principle leads to premise (2) of the argument from inductive risk, provided that acceptance is a decision that can have consequences for practical or ethical matters and can be made with regard to claims that are less than certain.[3]

This suggests a neat categorization of three possible ways to reject premise (2). One might (a) take issue with the consequence principle, for instance, by claiming that science should be given a special dispensation to pursue knowledge for its own sake and without regard for consequences. I will not examine this idea here as it has already been extensively critiqued (see Douglas 2009, chapter 4; Kitcher 2001, 2011; Kourany 2010). My focus instead will be on two other potential escape routes: (b) one could claim that acceptance decisions cannot (or should not) have consequences for practical actions or (c) one could insist that acceptance is reasonable only when one is sufficiently certain as to be unconcerned about error. These two possible lines of argument are considered in the next section.

[3] Cohen's is not the only theory of acceptance that possesses these characteristics (see Bratman 1999; Harsanyi 1985).

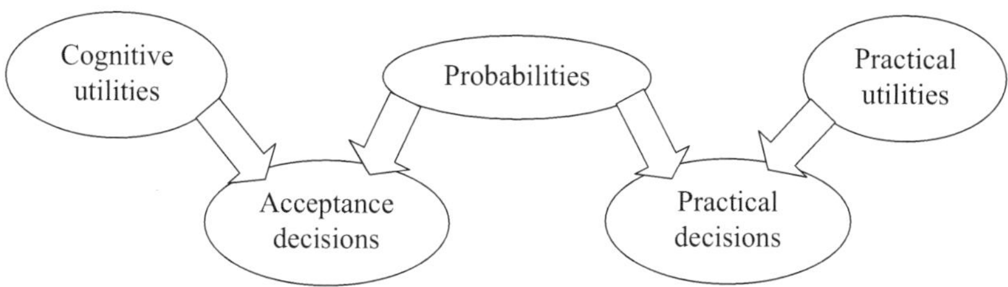


Figure 7.1 Cognitive utility theories of acceptance claim that, although our probability judgments affect both our decisions about practical matters and decisions about what to accept, what we accept should have no influence on practical decisions.

### 7.3.2 *Cognitive utilities and certainty*

In this section, I examine two groups of alternative theories of acceptance: one which explicates rational acceptance in terms of maximizing expected cognitive utility and the other which regards certainty in one form or another as a criterion of acceptance. These two approaches to acceptance did not arise independently of the argument from inductive risk. Significant representatives of both – particularly Levi (1967) and Lacey (1999) – were developed expressly for the purpose of countering Rudner's (1953) argument and explaining how acceptance of scientific claims can be grounded exclusively in epistemic values. Both of these approaches deny premise (2) of the argument from inductive risk, attempting escape by routes (b) and (c), respectively. I argue that both of these proposals suffer from serious difficulties not faced by Cohen's theory.

Consider cognitive utility theories. Whereas practical utilities are positive or negative valuations associated with overt actions, cognitive utilities are associated with acceptance. Thus, a dislike for bicycling in the rain and a zest for bicycling in the sunshine are matters of practical utility, while finding intrinsic value in discovering new knowledge is a matter of cognitive utility. According to cognitive utility approaches, moreover, maximizing expected utility is the standard for judging the rationality of decisions. Since expected utility is a probability weighted average of utilities, the bifurcation of cognitive and practical utilities results in two separate types of expected utility calculations, both sharing a common probability distribution but one using a cognitive utility function and the other a practical utility function (Harsanyi 1983; Levi 1967; Maher 1993). This set-up is depicted in Figure 7.1.

In Figure 7.1 there is no arrow pointing from acceptance to practical decisions or indeed to anything else. This is a deliberate feature of cognitive utility theories: they claim that what we accept should make no difference to decisions about practical action (Maher 1993, p. 150). In Levi, this move explicitly arises in response to the argument from inductive risk as advocated by Rudner (Levi 1960, 1967).[4] Advocates of cognitive utility theories do not regard acceptance as totally inert, however, because they hold that acceptance should satisfy deductive cogency (Levi 1967, p. 26; Maher 1993, p. 139). Deductive cogency states that if one accepts a set of sentences $A$ and is aware that $B$ is a logical consequence of $A$, then one ought to accept $B$ as well or else give up accepting $A$. But given that one can accept normative as well as descriptive statements, deductive cogency conflicts with the claim that what one accepts should have no implications for practical reasoning. Consider the following example. Suppose I accept that the glass of water before me contains arsenic at a level greater than 50 parts per billion ($W$). Suppose that I also accept the normative claim that if the glass of water contains arsenic at a level greater than 50 parts per billion, then I should not drink it ($W \rightarrow \neg D$). Obviously, $W$ and $W \rightarrow \neg D$ together logically entail $\neg D$ – in other words, that I should not drink the water. Thus, deductive cogency requires that I accept $\neg D$, assuming that awareness of this logical consequence does not make me doubt the reasonableness of accepting $W$ or $W \rightarrow \neg D$. This example illustrates that we can accept normative and descriptive claims that jointly entail further claims about what to do. Given deductive cogency, therefore, it is inescapable that descriptive claims we accept can function as premises in practical reasoning. But if this is so, how can it be that what we accept should have no impact on our actions, as cognitive utility theories assert? Perhaps advocates of cognitive utility theories would deny that we can accept normative statements. Yet this would be a highly implausible position, since people frequently do accept normative claims, for instance, that racism is bad (see Cohen 1992, p. 15). Moreover, Maher defines acceptance of a statement as the mental state expressed by the sincere and deliberate assertion of it (1993, p. 132). This apparently allows for the acceptance of normative statements, as people often do make sincere and deliberate assertions about what ought to be done. So denying that one can accept normative claims is not a promising option. What I think advocates of cognitive utility theories would say is that logical inferences from accepted descriptive and normative claims should not be a standard for rational action. Instead, they would claim, maximizing expected practical utility

[4] Maher takes for granted that practical consequences of error should not influence judgments about what to accept and argues on this basis that what we accept should not influence practical decisions (1993, p. 150).

is the one and only criterion for what it is rational to do (Maher 1993, pp. 149–50).

Yet the notion that logical consequences flowing from descriptive and normative claims one accepts should not influence one's actions is surely quite strange. For example, this would entail that practical and ethical reasoning framed in terms of logical inferences from accepted premises is irrelevant to human conduct. A further example may help to make vivid the deeply problematic nature of this position. Consider this famous argument proposed by Peter Singer (1972).

(a) Dying from lack of food, shelter, or medical care is very bad.
(b) If it is in our power to prevent something very bad without thereby sacrificing anything morally significant, we should do it.
(c) It is in our power to prevent people in poorer nations, such as Bangladesh, from dying from lack of food, shelter, or medical care at a small financial cost to ourselves.
(d) A small financial cost is not morally significant.
(e) Therefore, we should prevent people in poorer nations from dying from lack of food, shelter, and medical care.

Discussions of this argument typically presume that accepting its conclusion should have some impact on our behavior, for example, causing us to support organizations such as Oxfam International or Médecins Sans Frontières. If that is right, then whether or not one accepts the descriptive premise (c) could impact one's actions. But, according to cognitive utility theories of acceptance, this presumption is mistaken because, they insist, what we accept should make no difference to what we do. As Maher puts it, "*In any situation*, acceptance of *H* is consistent both with acting as if *H* were true and also with acting as if *H* were false" (1993, pp. 149–50, italics added). Thus, according to cognitive utility theories, a rational person could accept premise (c) of Singer's argument, along with the other premises, but deny that this constituted any reason to provide aid on the grounds that accepting (c) is consistent with acting as if it were false. Yet blithely disregarding the implications of what one accepts for practical action is unreasonable and perhaps even irrational. Nor can this be claimed as an unavoidable consequence of a Bayesian approach to decision, since some Bayesians have proposed theories of acceptance that do not have this undesirable feature (see Harsanyi 1985).

Attempts to use cognitive utility theories of acceptance as a basis for rejecting premise (2) of the argument from inductive risk, then, face the following dilemma. To say that accepted claims *should not* influence action is very problematic, given that it is possible to accept normative as well as descriptive claims and that the conjunction of the two can entail further claims about what ought to be done. On the other hand, if accepted claims *may* influence action, then there is a powerful argument for concluding – on the basis of the consequence principle as explained in section 7.3.1 – that practical and ethical factors can matter to decisions about what to accept.[5] In either case, cognitive utility theories fail to provide a plausible account of acceptance that undermines the argument from inductive risk.

Turn then to theories of acceptance that link acceptance to certainty. The second premise of the argument of inductive risk presupposes that a decision to accept a scientific hypothesis generally involves a non-negligible risk of error. Thus, some critics respond to the argument by claiming that acceptance, in a truly cognitive sense, requires a level of certainty that puts inductive risk out of the picture (Betz 2013; Dorato 2004, pp. 73–5; Lacey 1999, pp. 71–4; Levi 1980). For instance, Levi (1980) proposes that accepting a claim as part of our body of knowledge requires judging it to be completely certain. A central function of knowledge, Levi suggests, is to distinguish what are serious possibilities at a given time from what are not (1980, pp. 2–5). A claim is a serious possibility at a time just in case it is logically consistent with what we know then. Moreover, Levi says that only serious possibilities have a probability greater than zero (1980, p. 3). Hence, it follows from Levi's proposal that one must regard everything one accepts as knowledge as having probability 1, that is, as being absolutely certain.[6] Levi coins the term "epistemological infallibilism" to label this consequence of his proposal. Epistemological infallibilism would appear, on the face of it, to preclude the possibility that a person could envision discovering that something she accepts is false. However, Levi insists that a rational person should, despite being absolutely certain of the complete truth of all she accepts as knowledge, admit that some of what she accepts may be shown false by further inquiry (1980, p. 19). In other words, Levi claims that epistemological infallibilism is compatible with rejecting dogmatism and recognizing the corrigibility of what one accepts as knowledge.

Yet Levi's combination of infallibilism and corrigibility is paradoxical. If I am absolutely certain that a claim is true, then rationality requires that

[5] Kaplan (1981) proposes a Bayesian theory of acceptance not founded on the concept of maximizing expected cognitive utility but which nevertheless requires that what we accept should have no impact on action (Kaplan 1981, p. 330). Consequently, his proposal is also subject to this dilemma.

[6] Let $K$ be our corpus of knowledge at $t$. Then $\neg K$ is not a serious possibility at $t$ and hence $p(\neg K) = 0$. Thus, $p(K) = 1$.

I am similarly certain that all new truths discovered in the future will be logically consistent with it. And if I admit that a claim may be disproven in the future, I am expressing uncertainty about its truth. One cannot, in short, have it both ways: either you grant that your knowledge is fallible, and hence less than completely certain, or you insist that your knowledge is absolutely certain and thereby imply that it is infallible (Harsanyi 1985, p. 22). As a result, Levi has not provided a good explanation of how his position avoids tying acceptance to dogmatism.

Lacey also links acceptance to certainty, although, unlike Levi, does not regard complete certainty as a prerequisite for acceptance. Instead, he suggests that only "pragmatic" or "practical certitude" and not "epistemic certitude" is necessary (1999, pp. 14, 71). By practical or pragmatic certitude, Lacey means that the theory "is sufficiently well supported that it need not be submitted to further investigation – where, for example, it is judged that further investigation can be expected only to replicate what has already been replicated many times over and to bring minor refinements of accuracy" (Lacey 1999, p. 13). But Lacey's retreat from epistemic to practical certainty undermines his objection to the argument from inductive risk. For decisions about whether to accept a scientific claim in Lacey's sense involve a consideration of how probable or well supported it must be to justify the decision to no longer subject it to further investigation and to treat it as a closed case. Decisions of this kind can have practical consequences in a world in which scientific knowledge impacts decision making on matters of important public concern, such as climate change, regulation of toxic chemicals, or medical practices. The consequence principle can then be used to argue, as explained in section 7.3.1, that ethical or practical concerns can legitimately influence decisions about what to accept. Therefore, Lacey has provided no reason to undercut the argument that practical and ethical consequences of errors can be relevant to acceptance.

In addition, Lacey's proposal is problematic considered on its own terms because it links acceptance to a dogmatic dismissal of the usefulness of further inquiry. Yet accepting *S* does not require thinking that further research on *S* is pointless. For example, a scientist might accept that anthropogenic greenhouse gas emissions are the primary factor driving climate change while regarding further documentation of that claim as worthwhile. Indeed, she might even engage in such research herself. She could, for instance, judge this research to be useful because it can produce evidence that reduces uncertainty, which may be relevant to her own intellectual curiosity and which may be helpful for persuading others.

### *7.3.3 A hidden contradiction?*

As illustrated in the previous section, most critics of the argument from inductive risk grant premise (1) and challenge premise (2). That is, they agree that accepting and rejecting hypotheses is a proper part of science but claim that such decisions should, for the purposes of science, be construed in a purely cognitive manner that is unsullied by non-epistemic values. A less popular objection, classically made by Richard Jeffrey (1956), does the opposite. Jeffrey attempts to show that the premises of the argument from inductive risk lead to a contradiction, and since he agrees with Rudner that acceptance decisions depend on value judgments of an ethical or practical nature, Jeffrey concludes that (1) must be false.

Jeffrey's argument that accepting and rejecting hypotheses is not the proper role of scientists is based on the following two premises:

A. *if* it is the job of the scientist to accept and reject hypotheses, *then* he must make value judgments. (Jeffrey 1956, p. 237)

This is a restatement of premise (2) of the argument from inductive risk. Jeffrey then suggests the following premise:

B. if the scientist makes value judgments, then he neither accepts nor rejects hypotheses. (Jeffrey 1956, p. 237)

Given these two premises, the supposition that scientists should accept or reject hypotheses yields a contradiction, and, thus, by *reductio ad absurdum*, must be false. Jeffrey defends premise B on the grounds that a scientist could not foresee all of the purposes to which her hypothesis might be put. He supports his reasoning with an example of a scientist deciding whether to accept that a batch of polio vaccine is free from active polio virus. The scientist does not know if the vaccine is to be used for human or veterinary purposes, but a higher standard of evidence would presumably be demanded in the former case than the latter (Jeffrey 1956, pp. 241–2). Since the scientist does not know the intended application of the vaccine, she cannot know the consequences of accepting the hypothesis when it is false or of rejecting it when it is true. Hence, Jeffrey argues, the scientist is unable to use any rule for decision making under uncertainty – such as maximizing expected utility or the minimax rule – to decide whether a given body of evidence is sufficient to accept the hypothesis (Jeffrey 1956, pp. 240–5).

Rather than accept or reject hypotheses, Jeffrey proposed that scientists should simply report probabilities or degrees of confirmation, an idea echoed by several other philosophers who defend a Bayesian approach (Earman 1992, pp. 191–2; Howson and Urbach 1993, pp. 203–6). Defenders

Table 7.1 *The possible states in Jeffrey's polio vaccine example*

| | Human use & *H* is true | Human use & *H* is false | Veterinary use & *H* is true | Veterinary use & *H* is false |
|---|---|---|---|---|
| Accept *H* | | | | |
| Reject *H* | | | | |

of the argument from inductive risk often respond that claims about probabilities would also have to be accepted or rejected (Douglas 2009, pp. 53–4; Rudner 1953, p. 4). As a result, they suggest, Jeffrey's proposal does not avoid the argument from inductive risk. Although I agree with this point, it does not address Jeffrey's *reductio ad absurdum*. In other words, identifying a flaw in Jeffrey's positive proposal does not show that his case against the argument from inductive risk is mistaken.

One who defends the argument from inductive risk clearly must take issue with premise B, since premise A is shared by both arguments. And indeed, Jeffrey's defense of premise B is rather weak. To begin with, Jeffrey has not shown that familiar rules for decision making are inapplicable in the sorts of cases he describes. In his polio example, there two possible actions, accept or reject the hypothesis, and there are four possible states for each depending on whether the batch of vaccine is intended for human or veterinary use and whether the hypothesis is true or false (see Table 7.1). So if probabilities and utilities can be associated with each of these states, then some simple computations could determine whether accepting or rejecting the hypothesis maximizes expected utility. Perhaps Jeffrey's point is that decisions about acceptance might be made in situations involving uncertainty about the probabilities or utilities associated with different possible states. But uncertainty regarding probabilities and utilities may arise in decisions of all kinds and is not specifically linked to decisions about acceptance. Thus, the fact that decisions must sometimes be made in contexts of uncertainty or ignorance is a motivation for work in decision theory (see Peterson 2009; Resnik 1987), not an argument against making difficult choices.

Perhaps, then, Jeffrey's point is not that no rule of rational decision making can guide the scientist in the case he imagines but rather that there is a rule and it directs the scientist to leave the decision about what to accept to somebody else. That is, the scientist lacks crucial information about how the hypothesis will be applied, while people responsible for applying the hypotheses presumably possess this information and, consequently, know more about likely consequences of error. Thus, if one grants that those who possess the best relevant information should make the decisions, it might appear to follow that the scientist should not accept or reject hypotheses. Of the possible interpretations canvassed so far, I think this one results in the most plausible rendering of Jeffrey's argument. However, the argument so construed is nevertheless quite problematic for two reasons. First, it implicitly assumes that acceptance decisions must be the responsibility of a single person or homogeneous group of people (e.g., "the scientist," "the physician," or "the veterinarian"[7]) rather than the result of collaboration among individuals with distinct areas of expertise. Second, it overlooks the important fact that value judgments can be relevant to highly technical methodological or modeling assumptions that require scientific expertise to understand (Douglas 2000, 2009). Thus, even on this last interpretation, Jeffrey's argument does not provide good grounds for exempting or excluding scientists from acceptance decisions and the value judgments they entail. The sensible conclusion to draw is that scientists, possibly along with others, should play an active role in the process.

However, Jeffrey's objection does suggest a restriction of the scope of the argument from inductive risk, namely that argument is not pertinent when the consequences of acceptance or rejection are entirely unforeseeable. This restriction is supported by Douglas's account of the moral responsibilities of scientists: "they [scientists] are responsible for the foreseeable consequences of their choices, whether intended or not. Thus, scientists have the responsibility to be neither reckless nor negligent in their choices" (Douglas 2009, p. 71). In cases wherein the ethical or moral consequences of acceptance cannot be foreseen, those consequences would not be relevant to the decision. The argument from inductive risk is significant, however, because it often is the case that some knowledge exists regarding the consequences of errors.

## 7.4 Rethinking values in science

Abandoning the ideal of value-free science opens up space for an epistemic interpretation of PP, while calling for a rethinking of the difference between acceptable and unacceptable roles of values in science. My proposal for how to carry out this reconceptualization depends upon upholding the distinction between epistemic and non-epistemic values. Section 7.4.1 elaborates

[7] See Jeffrey (1956, p. 245).

the distinction, while sections 7.4.2 and 7.4.3 defend it against a number of objections.

### *7.4.1 Epistemic versus non-epistemic values*

I understand epistemic values to be values that promote truth-seeking aims (Goldman 1999). A number of qualifications are necessary to properly understand this proposal. First, truth should be understood in connection with truth content: a true and very informative claim is more epistemically valuable than a true but trivial one. This is important because it entails that an excessively cautious approach to drawing inferences can be defective from an epistemic standpoint. The rule that one should accept only what has been proven beyond all possible doubt, for instance, would obstruct the aim of acquiring informative truths and hence would not express a genuine epistemic value. Second, representations in science may be used for many different purposes and are often not intended to be truthful or accurate in all respects. For example, scientific representations are often models that include simplifying or idealizing assumptions known not to be strictly true. In an epistemic assessment of a scientific representation, then, it can be important to pay heed to its intended purpose and, consequently, to which aspects of it are intended to be truthful and which are not. Is the model intended for predictive purposes only, or to represent certain aspects of a causal process but not others, and at what grain or resolution of detail? Finally, the truth that is aimed for is not necessarily all or nothing, and in many cases "true enough" may be the intended target (Elgin 2004). This point is especially clear with respect to predictive models. Such models rarely generate forecasts with perfect accuracy, but it is not too difficult to understand what it is for a prediction of, say, the average global sea level in 2100 to be closer to or further from the truth. The concept of approximate truth has been discussed by a number of authors in the literature on scientific realism (e.g., Barrett 2008; Hunt 2011). However, the approach advanced here does not depend on the specifics of any of these proposals. But it is relevant for the present discussion to observe that decisions about what should count as "close enough" to true may depend on value judgments of an ethical or practical nature. I take that to be an illustration of the argument from inductive risk, because questions about what is good or close enough for some intended purpose are often elements of decisions about what should be regarded as sufficient evidence for accepting a claim.

Epistemic values, then, focus on factors that promote the attainment of truth-seeking aims without necessarily speaking to questions about why truths on that particular topic or to that degree of approximation are sought. A value might be truth promoting either intrinsically or extrinsically. Intrinsically valuable things are good in their own right, whereas extrinsic goods are valuable because they are means for promoting intrinsic goods (Anderson 1993, pp. 19–20). Thus, a value is intrinsically epistemic if exemplifying that value either constitutes an attainment of truth or is a necessary condition for a statement to be true. For example, predictive accuracy is an intrinsic epistemic value, because to be able to make accurate predictions is to have attained an important truth-seeking aim, namely the acceptance of true or approximately true statements about the future. The epistemic status of consistency is somewhat more complex. Consistency is commonly divided into two subtypes: internal and external. Internal consistency means an absence of self-contradictions, while external consistency refers to an absence of contradictions with other accepted claims. Internal consistency is a straightforwardly intrinsic epistemic value, since it is a necessary condition for truth.[8] Whether external consistency is an epistemic value, on the other hand, depends on the truthfulness of the accepted background claims. If those claims are significantly false (e.g., that the Earth is the center of the universe and no more than 10,000 years old), then a demand for external consistency can be a major epistemic impediment. However, when external consistency is an epistemic value (i.e., when one's background knowledge is largely accurate), it is an intrinsic epistemic value for the same reason that internal consistency is: if a sentence *S* is true, then consistency with *S* is a necessary condition for truth.

Epistemic values are extrinsic when they promote the attainment of truth without themselves being indicators or requirements of truth. A good example of an extrinsic epistemic value is Popper's criterion of testability. The more precise and informative a theory's empirical predictions are, the greater its testability. Testability alone is not an intrinsic epistemic value, because a theory that makes a wide range of precise yet false predictions is nevertheless very testable and an untestable theory might be true. However, Popper argued that a preference for testable theories promotes the attainment of truths by increasing the efficiency of scientific inquiry (1963). An untestable theory, even if true, is a scientific dead end that leads to no further gains in empirical knowledge, while highly testable theories, even if false, generate predictions that may spur scientific discoveries.

As another example of an arguably extrinsic epistemic value, consider simplicity. Simplicity would be an intrinsic epistemic value only if the

[8] I assume here that contradictions are false, contrary to dialetheism (see Priest 1998).

world were simple, since otherwise there would be no reason to judge simplicity as an indicator of or requirement for truth. Yet a general presumption that we inhabit a simple world hardly seems a promising basis for simplicity as an important epistemic value. But simplicity might be an extrinsic epistemic value even if it is not an intrinsic one. There are in fact two approaches for making this case with regard to examples, such as fitting curves to a scatter of data, in which simplicity can be given a reasonably precise meaning. One approach is to show (subject to various qualifications) that minimizing expected predictive errors requires selecting a curve that strikes the right balance between simplicity and closeness of fit with the data (Forster and Sober 1994). According to this proposal, while selecting a more complex curve can normally produce a closer fit with data, "over fitting" the data with an excessively complex curve increases the chance of a greater mismatch between future data and predictions. A second account of simplicity explains how a preference for simpler hypotheses promotes efficient convergence to the truth, where efficiency is understood in terms of minimizing the maximum number of times the investigator can switch from conjecturing one hypothesis to another (Kelly 2007a, 2007b; Steel 2009). Both of these accounts, then, defend simplicity as an extrinsic epistemic value. In both cases, a preference for simpler hypotheses is argued to promote the attainment of truth (either of approximately true predictions or efficient convergence to true hypotheses), yet neither approach presumes that the world is simple.

There are some other aspects of epistemic values as understood here that are worth making note of. Things other than theories and hypotheses, such as methods, social practices, and community structures, can manifest epistemic values. For example, several philosophers have proposed that open debate among peers who represent a diversity of viewpoints is more likely to promote the acquisition of truth than a hierarchical system in which investigators must defer to authorities (Longino 1990; Popper 1966, pp. 217–18). If that is right, then values that promote free and open discussion in science are epistemic values. Similarly, if the practice of granting credit for a scientific discovery to the first person to make it has a positive impact on the truth-finding ability of science, as some have suggested (Kitcher 1990; Strevens 2003), then it, too, would count as an epistemic value.

Another important feature of epistemic values is that they often interact with one another. For example, testability *alone* is not an intrinsic epistemic value, but testability is an intrinsic epistemic value when combined with empirical accuracy. For theories that are empirically accurate, greater testability means more informative true empirical consequences. But testability is not an intrinsic epistemic value for highly inaccurate theories: for such theories, greater testability merely translates into more precise falsehoods. In addition, some epistemic values, such as empirical accuracy, are very robust in the sense of being epistemic in an almost any setting. Other epistemic values are contextual in the sense that their capacity to promote the attainment of truth depends on occurring within a specific set of circumstances. That point is illustrated by external consistency: its status as epistemic depends on the truthfulness of what is taken as established knowledge. Contextual epistemic values also raise the possibility that a value might be epistemic in one historical period but not in another. For example, external consistency might fail to be an epistemic value in a period in which background assumptions are seriously mistaken but become an epistemic value at a later time when these errors have been corrected.

It should be emphasized that the distinction between intrinsic and extrinsic is independent of the distinction between robust and contingent. All four combinations – intrinsic robust, intrinsic contingent, extrinsic robust, and extrinsic contingent – are possible. Empirical accuracy is an example of an intrinsic and robust epistemic value, while the status of external consistency as epistemic is contingent but when it is an epistemic value it is an intrinsic one. Testability is a good candidate for a robust yet extrinsic epistemic value: it is extrinsic for the reasons noted above, but the argument for preferring testable hypotheses appears applicable to nearly all branches of science. Finally, if a rule bestowing credit for scientific breakthroughs solely to the initial discoverer is an epistemic value, it is extrinsic and, I suspect, also contingent.

As drawn here, the distinction between epistemic and non-epistemic values differs from a number of other distinctions with which it is often associated. For instance, the distinction between epistemic and non-epistemic values does not correspond to a distinction between facts and values. Epistemic values are not distinguished by being factual rather than evaluative: they are distinctive in virtue of promoting the acquisition of truths. The difference between epistemic and non-epistemic values is also not the same as Longino's distinction between constitutive and contextual values (1990, pp. 4–7). Constitutive values are grounded in the aims of science itself, while contextual values are features of the broader social environment of which science is a part. This is not the same as the epistemic/non-epistemic distinction, because a contextual value might promote the attainment of truths and a constitutive value might fail to do this if truth is not the sole aim of science.

A word on the concept of truth assumed in this discussion is in order. All that I assume about truth is that it satisfies the following transparency condition: the sentence "P" is true if and only if P. For example, "Ice is cold" is true if and only if ice is cold. Despite its seemingly trivial nature, the transparency condition does conflict with some theories of truth. For instance, consider a pragmatic theory of truth according to which whatever is useful for a person to believe is true. That this pragmatic theory of truth violates the transparency condition can be seen by examples such as the following: it is useful for Tom, who is 5′10″, to believe that he is six feet tall, since this belief enhances his self-esteem. There are several theories of truth that satisfy the transparency condition, and for the present purposes there is no need to express favor for any one of them over the others.[9] Finally, I should say something about how the term "value" is used here. Values function as what Miriam Solomon calls "decision vectors," that is, factors that "influence the outcome (direction) of a decision" (2001, p. 53). However, not all decision vectors would normally be thought of as values. For example, sexist bias is a decision vector but would not be a value in a social setting in which sexism is frowned upon. Values, then, are decision vectors that are favorably regarded in a community.

The next two sections consider a number of objections to the epistemic/non-epistemic distinction. Some of these objections question the distinction generally, while others raise concerns about specific features of the distinction proposed here. I begin with objections of the former type.

### 7.4.2 *Defending the distinction*

One line of objection claims that there is no clear distinction between epistemic and non-epistemic values. For example, Phyllis Rooney argues that the lack of agreement among proposed lists of epistemic values provides evidence of this: "The fact that there is no consensus about what exactly the epistemic values are surely provides our first clue here. We haven't seen anything resembling a clear demarcation of epistemic values because there is none to be had" (1992, p. 15).

But disagreement about the application of a concept is not necessarily evidence that it lacks a reasonably clear definition. For example, spectators at soccer matches sometimes disagree about whether a goal was scored on a particular play. But this is not evidence that there is no clear definition of

[9] See Blackburn and Simmons (1999) for a survey of the leading theories.

what constitutes a goal in the game of soccer (the criterion is that the entire ball must cross the goal line). Rather, disagreements result from the fact that, in some cases, it is difficult to discern whether the definition is satisfied, as illustrated by Geoff Hurst's famous – or to German soccer fans, infamous – game-winning goal in the 1966 World Cup Final. Similarly, disagreements about epistemic values are not necessarily evidence that there is no clear definition of epistemic. According to the account of epistemic values proposed here, disagreements occur because it is sometimes a very complex and difficult question whether or not a particular value promotes the attainment of truth. Moreover, my proposal can explain why some values are regarded by almost everyone as epistemic while other values are more controversial. Every list of epistemic values that I know prominently includes empirical accuracy or some close variant. Since empirical accuracy is a robust and intrinsic epistemic value, this is exactly what one would expect given my approach. On the other hand, the status of simplicity as an epistemic value is far more contentious. That is also not surprising. Simplicity is usually not an intrinsic epistemic value, and attempts to show that simplicity is an extrinsic epistemic value require a heavy dose of technical concepts and complex mathematical or logical reasoning. So, the uncontroversial epistemic values are obviously truth conducive, while the controversial epistemic values are ones whose connection to truth is much more complex. This situation is more easily explained by the approach defended here than by the claim that there is no clear distinction whatever between epistemic and non-epistemic.

Another argument that there is no clear distinction between epistemic and non-epistemic values points out that it is impossible to disentangle social from epistemic (Machamer and Douglas 1999; Machamer and Ozbeck 2004). I think that this line of criticism has merit as an objection to some proposals about how to distinguish epistemic from non-epistemic values,[10] but it is not an objection to the account advanced here. Given my account, social is not the appropriate contrast with epistemic. Epistemic values are contrasted with values that do not promote the attainment of truth, either because they have no impact whatever on that goal or because they obstruct it. Both epistemic and non-epistemic values can be and typically are social. Indeed, an epistemic value must be embedded in the social practices and norms of a scientific community if it is to have any influence on science. Epistemic, therefore, should not be contrasted with

[10] For example, see Machamer and Douglas's (1999) critique of Lacey's attempt to distinguish cognitive values from social values by means of "materialist strategies" (1999, p. 68).

social, historical, contingent, or contextual. An epistemic value may be a social norm with a complex history, and its status as epistemic may depend on contingent facts about the world and the context of social practices within which it operates. Sometimes the thrust of the can't-disentangle-social-from-epistemic objection seems to be that the non-epistemic values prevalent in a social milieu can influence what a person thinks the epistemic values are (Douglas 2009, p. 90; Rooney 1992). It seems very probable that this claim is true, but that is no reason for concluding that there is no reasonably clear distinction between epistemic and non-epistemic values. To continue the soccer analogy, a person's allegiances may influence whether she thinks a goal was scored (e.g., English fans think Hurst's goal was genuine while German fans do not), but this does not show that the term "goal" lacks a clear definition in the game of soccer.

Another objection is that epistemic values are too anemic to guide scientific inquiry toward consensus and, consequently, that other sorts of values must influence scientific judgments (Douglas 2009, pp. 93–4; Laudan 2004; Longino 1996; Rooney 1992, pp. 15–20). This objection is a challenge to the usefulness of the concept of epistemic values rather than to the cogency or existence of that concept. The thought behind the objection seems to be that defenders of epistemic values believe that, in a perfect world, epistemic values alone would guide scientific inference. Hence, the objection goes, if epistemic values can be shown incapable of this, the reason for distinguishing epistemic from non-epistemic vanishes and the distinction is useless. However, the objection shows, at best, only that the distinction between epistemic and non-epistemic values is not useful for defending the ideal of value-free science. But I think it should be clear that this is not the only use to which the epistemic/non-epistemic distinction might be put. My purpose in upholding that distinction is to provide a better defense and articulation of the argument from inductive risk, which attempts to show that non-epistemic values *should* influence decisions to accept or reject hypotheses. In addition, I think these critics tend to exaggerate the incapacity of epistemic values to guide scientific inference. For instance, they often do not consider the possibility of extrinsic epistemic values. Thus, Helen Longino insists that simplicity could be an epistemic value only if the world were simple (1996, p. 53). Another mistake is to neglect the difference between truth content and probability. From an epistemic perspective, the more truth the better, and hence it is epistemically valuable for an empirically accurate theory to be highly informative regardless of whether it is highly probable. Larry Laudan misses this point when he suggests that there can be no epistemic reason for saying that a theory is good but false and no epistemic grounds judging a theory with greater scope to be superior to one that is more narrowly restricted (2004, pp. 18–19). After all, Laudan reasons, the probability of the false theory is zero, and the probability of the more restricted theory must be greater than or equal to the probability of the one with greater scope. But this only shows that probability is not the same as truth content. A false theory may be of epistemic value if it makes many true predictions (as the case of Newtonian mechanics illustrates), and the wider the range of accurate predictions the better.

### *7.4.3 In defense of contingent epistemic values*

So far I have considered general objections to the distinction between epistemic and non-epistemic values. In this section, I consider an objection to a specific feature of the epistemic/non-epistemic distinction drawn here. Sven Diekmann and Martin Peterson (2013) object to the idea that whether a value is epistemic can depend upon contingent facts about the world:

> Arguably, which values count as epistemic should not depend on contingent facts that vary from one context to another. The purpose of grouping values as epistemic or non-epistemic is to distinguish those values that are, at a quite general level, beneficial from an epistemic point of view from those that are not.
>
> What counts as a (certain type of) value does not vary from one context to another. Values do not behave like that. (2013, p. 210)

According to Diekmann and Martin, then, a contingent value could not provide context-independent epistemic guidance and hence could not be properly judged epistemic. Translated into the terminology proposed in the previous section, therefore, their argument rests on the premise that only robust epistemic values should truly count as such. However, this argument is problematic in several ways. Some commonly cited examples of epistemic values are clearly context dependent. As noted above, whether external consistency is an epistemic value depends on the accuracy of the accepted principles in one's field, which surely does vary across contexts. A similar point could be made about simplicity. For instance, Malcom Forster and Elliott Sober's (1994) defense of simplicity requires that the space of possible hypotheses have a specific nested structure. When this condition is not met, the defense of simplicity they propose would not be relevant and spurious invocations of simplicity as an epistemic value might arise. Thus, as a descriptive matter, the use of the term "epistemic value" is not restricted in the manner that Diekmann and Martin recommend.

Moreover, there are good reasons for not narrowing the definition of "epistemic value" in this way. For consider a scientific field in which accepted background principles are very accurate: in such a field, external consistency could be appropriately cited as an important epistemic value. Yet it would, I submit, make little sense to deny that external consistency is an epistemic value in this field because there are *other* fields in which background assumptions are faulty. Similarly, experience in a particular field may have revealed that a certain experimental procedure is extremely important for achieving reliable results. A norm that researchers follow this procedure could, therefore, be sensibly regarded as an epistemic value in that discipline. But again it would be unreasonable to insist that the norm is not really epistemic in this context because there may be other fields of scientific research wherein it would be ineffective. In general, the idea that epistemic values must be context independent seems to be implicitly grounded in an implausible picture of scientific methodology according to which effective epistemic norms are somehow established independently of the accumulated experience of investigators. For if it is otherwise – that is, if the past experiences of researchers are relevant to knowing what the epistemic values are – then it is unavoidable that epistemic values can be contextual because what has proven epistemically beneficial in one domain may or may not produce similar results elsewhere.

Nevertheless, I think the distinction between robust and contingent epistemic values is useful and important. One reason is that robust epistemic values can identify general patterns of problematic reasoning that arise repeatedly in a variety of contexts. In such situations, an exclusive focus on domain-specific epistemic norms would cause one to miss the forest for the trees, which in turn could hinder efforts to effectively recognize and respond to a systemic problem. I believe that the case of "bent" (McGarrity and Wagner 2008) or "special interest" science (Shrader-Frechette 2011) illustrates a systemic problem of this kind. Special interest science is intended to achieve some legal or commercial function, such as getting a pharmaceutical approved by the US FDA or defense from a product liability lawsuit. The general pattern is that there is a certain hypothesis that the funder of the research would strongly prefer to see vindicated, and this preference is independent of the truth or accuracy of the hypothesis. For instance, the hypothesis might be that Vioxx does not have harmful side effects, or that second-hand smoke does not cause cancer. This situation can create a powerful economic incentive to generate studies that superficially resemble ordinary scientific research but in which the preferred hypothesis was never subject to any genuine risk of refutation. In short, special interest science often involves a violation of the epistemic value that test results taken as evidence for a hypothesis must put that hypothesis to a severe test (Mayo 1996; Popper 1963). The severity criterion plays an important role in the exposition of the values-in-science standard proposed in the next chapter.

## 7.5 Conclusions

The notion of precautionary science entails an epistemic interpretation of PP, which requires rejecting the ideal of value-free science and, consequently, rethinking the role of values in science. This chapter defended what I take to be the most significant objection to the value-free ideal, namely the argument from inductive risk. I explained how that argument is rendered more philosophically defensible when conjoined with a conception of accepting as premising due to Cohen (1992). But I also defended the traditional distinction between epistemic and non-epistemic values. This distinction is not only compatible with the argument from inductive risk but is also useful for reconceiving the relationship between science and values after the demise of the value-free ideal.

CHAPTER 8

# *Values, precaution, and uncertainty factors*

## 8.1 Case study needed

An abstract philosophical critique of the ideal of value-free science may leave one with the uncomfortable feeling that, while all this talk of values affecting judgments about which hypotheses scientists accept might sound nice in theory, it would never work in practice. In this chapter, then, I lead off with a detailed case study of a familiar procedure in toxicology in which just such a role of values is implicated. Toxicologists often rely on extrapolation from animal experiments to estimate human acceptable daily intake levels, or reference doses, for potentially toxic chemicals. Uncertainty factors, which work by dividing a level of exposure found acceptable in animal experiments by a number (often 100) to generate a reference dose, are a commonly used method in such extrapolations (Nielsen and Øvrebø 2008). Since uncertainty factors allow for the estimation of reference doses in the face of considerable uncertainties, and do so in a way that is intended to err on the side of protecting human health and the environment, they would seem to be a clear-cut illustration of an application of an epistemic PP.

After describing uncertainty factors in section 8.2, I propose a norm called the *values-in-science standard*, which is intended to delimit the appropriate roles that non-epistemic values may play when impacting scientific decisions about which hypotheses to accept or reject. The central idea is that non-epistemic values should not override epistemic values in feasible and ethically permissible scientific research. I explain how uncertainty factors are compatible with this standard, while special interest science often is not. In section 8.3.3, I compare the values-in-science standard to Douglas's proposal to replace the value-free ideal with an approach based upon a distinction between direct versus indirect roles of values. Despite some significant overlap, they differ insofar as the science-in-values standard insists that non-epistemic values in both roles are constrained by epistemic values. And in section 8.3.4, I consider objections that the values-in-science standard is either too permissive or too strict in the role it allows for non-epistemic values.

Finally, I tie the discussion of uncertainty factors and the values-in-science standard back to the epistemic PP. Two themes emerge here. First, uncertainty factors are an epistemic application of MPP, according to which policy-relevant science should avoid excessively slow and cautious scientific methods when confronted with urgent threats to the environment or human health. The second theme is that *which* risks are quantifiable and *how* they are quantified depends on what methods for quantifying risks are deemed scientifically legitimate, and *PP is relevant to deciding what those methods should be*. Which methods we choose for risk assessments depends in part on what priorities we place on avoiding which types of errors, and PP has something to say about such questions. A corollary of this second theme, then, is that it would be a mistake to restrict application of PP to unquantifiable risks.

## 8.2 Uncertainty factors

In this section, I begin by describing how uncertainty factors function in risk assessment. Next, I examine several arguments provided for justifying particular numerical values for uncertainty factors, going back to the seminal essay on the topic by A. J. Lehman and O. G. Fitzhugh published in 1954.

### *8.2.1 How uncertainty factors work*

Uncertainty factors are a commonly used device in the estimation of reference doses (Nielsen and Ørevbø 2008). A reference dose is defined by the US Environmental Protection Agency (EPA) as "an estimate, with uncertainty spanning perhaps an order of magnitude, of a daily oral exposure to the human population (including sensitive subgroups) that is likely to be without an appreciable risk of deleterious effects during a lifetime."[1] Reference doses are typically estimated from toxicological experiments on animals, and uncertainty factors play a role in this animal-to-human extrapolation. Uncertainty factors (also known as safety factors) were originally proposed in the 1950s for use by the US Food and Drug Administration (FDA) in association with no-observed-adverse-effect levels (NOAEL) (Lehman and Fitzhugh 1954). Animals in an experiment are exposed to several

[1] See the EPA's online glossary: http://www.epa.gov/risk_assessment/glossary.htm.

distinct dosage levels (typically measured in milligrams per kilograms of body weight), and the largest dose that exhibits no statistically significant harmful effect is the NOAEL. The NOAEL is then divided by the uncertainty factor to arrive at a reference dose that is regarded as safe for human exposure:

$$\text{Reference dose (RfD)} = \frac{\text{NOAEL}}{\text{Uncertainty factor}}$$

For example, if the NOAEL from animal experiments is 0.1 milligram per kilogram of body weight per day (mg/kg/day) and the uncertainty factor is 100, then the reference dose would be 0.001 mg/kg/day.

The uncertainty factor itself is broken down into several components each of which represents a source of uncertainty present in the estimation of the reference dose. The following is what might be considered the "classic" list of default uncertainty factors for reference doses (Dourson, Felter, and Robinson 1996; Dourson and Stara 1983):

- *Interspecies extrapolation*: This uncertainty factor takes account of the possibility that humans are more susceptible to the harmful effects of the substance than the animals (for instance, rats) in which the NOAEL was estimated. Its default value is 10.
- *Intra-species variation*: This uncertainty factor reflects the fact that there are typically subgroups of the population, such as infants and the elderly, who are more susceptible than average. (Recall that the reference dose is intended to be protective of sensitive subgroups.) The intra-species uncertainty factor is important because populations of laboratory animals are normally much more homogeneous than human populations. Its default value is 10.
- *Sub-chronic to chronic*: Chronic exposure is exposure occurring over an extended period of time. This uncertainty factor, then, is used when the only available data derive from studies that examined only the effects of short-term exposure to the substance. Its default value is 10.
- *LOAEL to NOAEL*: In some cases, animal experiments may fail to find a dose that does not produce an adverse effect. In this case, a lowest-observed-adverse-effect-level (LOAEL) is used as a basis for estimating the NOAEL, and an additional uncertainty factor with a default value of 10 is applied.
- *Modifying factor*: This is an uncertainty factor for residual uncertainties not covered by the above uncertainty factors, such as incomplete data or flawed research designs in the available studies. It may vary from 1 to 10.

The separate uncertainty factors are multiplied to generate the total. However, not every uncertainty factor applies in every case. For example, if methodologically sound, chronic studies in animals have found a clear NOAEL, then only the first two would be relevant, resulting in a total default uncertainty factor of $100 = 10 \times 10$. Uncertainty factors are a common and relatively uncontroversial aspect of risk assessment. For example, the classic default uncertainty factors listed above, or close variants of them, are used by the US EPA, Health Canada, and the World Health Organization's International Programme on Chemical Safety, among others (Dourson, Felter, and Robinson 1996).

Uncertainty factors are not limited to use in combination with NOAELs, which is a good thing given that there are a number of limitations inherent in NOAEL methodology (Filipsson *et al.* 2003). For example, since the NOAEL is an actually administered dose, the resulting reference dose is highly sensitive to which dosages happen to be chosen in the experiment. Furthermore, the approach does not take into account uncertainties inherent in estimating the NOAEL – in particular, that the absence of a statistically significant result might reflect a lack of statistical power of the test (e.g., because of small sample size) rather than the true absence of an effect. In addition, the NOAEL disregards information about the effects of other doses. Presumably, it should matter to risk assessment whether the adverse effects increase very sharply or only very gradually at doses higher than the NOAEL. Finally, the NOAEL approach is applicable only in cases in which there is a threshold below which no adverse effect occurs, which is generally thought not to be true of carcinogens.

There are more sophisticated approaches that are designed to avoid these shortcomings of NOAELs. Probably the most common of these is the benchmark dose (BMD) approach (Crump 1984; Filipsson *et al.* 2003; Kimmel and Gaylor 1988). In this approach, the goal is to identify the dosage (called the BMD) that raises the probability of an adverse effect by some specified small amount, usually a number between 1% and 10%. Rather than focusing on a single dose, the BMD approach estimates a dose-response curve that represents how the severity of the adverse effect rises as the dosage is increased. In addition, there is an element of randomness in any experiment to estimate a dose-response curve; if the experiment were to be repeated, it is very unlikely that the same estimate would be obtained each time. In order to account for uncertainties involved in the estimation of the dose-response curve, a confidence interval is also estimated around it. Think of two additional curves, one above and one below, equidistant from and running parallel to the original dose-response curve. The higher curve says that the substance is in fact more harmful than the data suggest, while the lower curve says that it is less harmful. The confidence interval, then, is the area between these upper and lower curves. The confidence interval is chosen so that, were the experiment to be repeatedly performed, the true curve would fall within it 95% of the time. The larger the sample size (i.e., the number of animals in the experiment), the narrower the 95% confidence interval. The BMD approach, then, errs on the side of caution by using the upper curve as the basis for estimating the dose of the chemical that would raise the probability of the adverse effect by the specified amount (say, 1%). This lower, more protective estimate is called the BMDL, where the "L" stands for "lower." Uncertainty factors can then be applied to the BMDL as they would to a NOAEL (the only exception being that the LOAEL to NOAEL uncertainty factor would not be relevant to a BMDL). The BMD method is widely used in risk assessment, for instance, by the US EPA.

### *8.2.2 Why* these *numbers?*

One obvious question to ask about uncertainty factors is what grounds there are for choosing one default rather than another. Why have 10 as the default instead of 3 or 30? I will approach this question by examining some of the attempts to justify uncertainty factors, beginning with Lehman and Fitzhugh (1954).

Rather than the list of specific uncertainty factors, each addressed to a distinct source of uncertainty, Lehman and Fitzhugh suggested a block "100-fold margin of safety" (1954). They provided several justifications for this proposal, some of which correspond to entries on the list of classic default uncertainty factors given above. Thus, they point out that laboratory animals are usually less sensitive to toxic chemicals than humans, a claim they support with several examples in which separate toxicity data exist for both. For example, humans are about 10 times more susceptible to the toxic effects of fluorine than rats, and about 4 times as sensitive to arsenic than dogs (1954, pp. 33–4). In addition, they note that populations of laboratory animals are typically more homogeneous than human populations (1954, p. 34). Lehman and Fitzhugh conclude: "The '100-fold margin of safety' is a good target but not an absolute yardstick as a measure of safety. There are no scientific or mathematical means by which we can arrive at an absolute value" (1954, pp. 34–5). Subsequent attempts to provide justifications for uncertainty factors have often preceded by breaking down Lehman and Fitzhugh's "100-fold margin of safety" into more fine-grained components, which then could be assessed more directly in relation to data.

One of the earliest proposals was to decompose the 100-fold margin of safety into 10 for interspecies variation and 10 for intra-species (Bigwood 1973, p. 70). Bigwood and Gérard (1968, pp. 218–20) considered three concerns relevant to the intra-species factor (differences in food requirements, differences in water intake, and differences in hormonal functions) and two relevant to the interspecies uncertainty factor (differences in body size between experimental animals and humans, and variation among experimental animals in susceptibility to toxic substances). They judged that each of the three intra-species considerations merited an uncertainty factor of 2, while judging the first interspecies consideration to merit an uncertainty factor of 8. The grounds for these judgments were not entirely clear, but they appear to have been that the numbers were sufficient to cover the known variation of each category. Bigwood described variations in susceptibility to toxins among animal species as "impossible to deal with quantitatively" but suggested that it probably could be accommodated by rounding the product of the other components ($2 \times 2 \times 2 \times 8 = 64$) up to 100 (1973, p. 71). The result, Bigwood suggested, was a roughly approximate, but "by no means arbitrary," default uncertainty factor (ibid.).

The idea that differences in body size between laboratory animals and humans support the interspecies uncertainty factor was further explored by Dourson and Stara, who interpreted it as the claim that the toxic effects of a chemical in distinct species is proportional to surface area (1983, pp. 228–30). As a matter of geometry, larger organisms generally have less surface area per body weight than smaller organisms. Thus, since dosage is measured by milligrams (mg) per kilograms (kg) of body weight per day, the proportionality of toxic effect to surface area entails that the same dosage level will be more toxic for larger than for smaller organisms. For example, suppose that the NOAEL of chemical X for a 1-kg guinea pig is .05 mg per day. Directly extrapolating this NOAEL to a 70-kg human without applying any uncertainty factor would result in a reference dose of .35 mg (= $70 \times .05$ mg) per day. But if toxic effect is proportional to surface area, a reference dose of .35 mg per day would not be protective for a 70-kg human, because the human would have between 4 and 5 times less surface area per body weight than the guinea pig. Thus, the .35 mg dose for a 70 kg human would mean 4 to 5 times more exposure per unit of surface area, and hence 4 to 5 times more toxic effect, than the .05-mg dose for a 1-kg guinea pig. However, an interspecies uncertainty factor of

10 would be sufficiently protective in this case. Dourson and Stara point out that if toxic effect is proportional to surface area, a default interspecies uncertainty factor of 10 is sufficiently protective when extrapolating from most, though not all, animals used in toxicology experiments – mice being the most important exception.

However, the assumption that toxic effects are constant per unit of surface area is rather crude. For instance, differences in susceptibility to the toxic effects of a chemical are often linked to differences in its metabolism or rate of excretion, which can vary independently of surface area. More recent approaches to providing an empirical basis for uncertainty factors, then, break the default of 10 for the inter- and intra-species uncertainty factors into separate components for toxicokinetic and toxicodynamic differences (Walton, Dorne, and Renwick 2001a, 2004). Toxicokinetics has to do with the mechanisms by which toxic substances enter and travel through the body. Toxicokinetics is divided into four sub-processes: absorption, distribution, biotransformation (or metabolism), and excretion. Metabolism involves chemically modifying the original substance, for example, by oxidizing it. Toxicodynamics, on the other hand, has to do with the mechanisms by which the toxin causes harm, for example, through provoking an immune response or damaging DNA. The interspecies uncertainty factor is split into $10^{0.6}$, or 4, for toxicokinetics and $10^{0.4}$, or 2.5, for toxicodynamics, while the intra-species uncertainty factor assigns $10^{0.5}$, or approximately 3.16, to both toxicokinetics and toxicodynamics (Dorne and Renwick 2005, p. 21). The two components could be modified on the basis of case-specific data, and the resulting values would be multiplied to derive the uncertainty factor. This approach is useful from the perspective of justifying uncertainty factors, because it is easier to obtain reliable human data on toxicokinetics and toxicodynamics than it is to get good human data on NOAELs. Moreover, assumptions connecting toxicokinetics and toxicodynamics to overall toxic effect rest on firmer scientific ground than, for instance, the assumption that toxic effect is proportional to surface area.

One approach pursued by toxicologists is to study the kinetics and dynamics of substances in the human body by way of pharmaceutical databases. Such studies rely on assumptions that connect toxicokinetics to toxic effect: for example, that quick removal of a chemical from the body is likely to reduce its propensity to have a toxic impact. Thus, it might be possible to show that chemicals of a certain class tend to persist in higher concentrations in the bloodstream of humans than in rats or in pregnant women than in other adults. Given quantified data of this kind, it would be possible to assess how protective the default uncertainty factors are. Studies of this sort have examined both the interspecies (animal to human) and intra-species (human variation) default uncertainty factors. Studies on the intra-species default uncertainty factor suggest that it is sufficient to protect nearly all of the healthy adult population, but that it is less protective for some important subgroups, including pregnant women and infants (Dorne, Walton, and Renwick 2001; Renwick and Lazarus 1998). Studies also suggest that the interspecies toxicokinetic default uncertainty factor of 4 would often be insufficient to cover differences in metabolism between humans and some commonly used laboratory animals, depending on the animal and the pathway through which the chemical is metabolized (Walton, Dorne, Renwick 2001a, 2001b). In fact, there are some metabolic pathways and animals for which the animal-to-human clearance ratio is greater than 10, that is, greater than the entire interspecies default (2001a, p. 676; 2001b, p. 1185). Such results undermine claims by critics that the default interspecies uncertainty factor of 10 is almost always excessive and should be replaced with a smaller number such as 3 (Chapman, Fairbrother, and Brown 1998, p. 106; Lewis, Lynch, and Nikiforov 1990, pp. 320–1).

## 8.3 Replacing the value-free ideal

In this section, I present and defend the values-in-science standard (section 8.3.1) and apply it to the case of uncertainty factors (section 8.3.2). In section 8.3.3, I consider an alternative proposal made by Douglas (2009) for how to delineate acceptable from unacceptable influences of values in science given that the value-free ideal is rejected. Finally, in section 8.3.4, I consider a number of objections to the values-in-science standard.

### *8.3.1 The values-in-science standard*

Given the distinction between epistemic and non-epistemic values, I propose that scientific integrity imposes the following constraint:

> **Values-in-science standard**: Non-epistemic values should not conflict with epistemic values in the design or interpretation of scientific research that is practically feasible and ethically permissible.

The values-in-science standard assumes that ethical restrictions may prohibit certain research designs (e.g., a randomized experiment in which humans subjects are deliberately exposed to a suspected carcinogen). I take it that this role of ethical values in science is uncontroversial. The values-in-science standard focuses on the role of values in decisions among research designs and interpretations of data that are practically feasible and do not contravene ethical principles concerning the treatment of research subjects. An example of such a decision might be the choice between distinct strains of rat in a toxicology experiment or the decision to base conclusions on one statistical analysis rather than another. By the design of scientific research, I mean decisions concerning methods and procedures that generate data – for example, in an experiment – while interpretation refers to the processes by which hypotheses are inferred from data generated by a study. Applications of the values-in-science standard obviously depend on how conflicts with epistemic values are defined. I will say that a conflict with epistemic values arises when a decision is recommended against by an epistemic value or rule that is unambiguously relevant to the case and is not counterbalanced by any other epistemic value.

To illustrate, consider an example of what I take to be a very robust epistemic value, namely the severity principle. In its simplest form, the severity principle asserts that a test counts as evidence for a hypothesis only if that test puts the hypothesis at risk of a contrary result (Mayo 1996; Popper 1963). That is, the severity principle states that "data $x_0$ do not provide good evidence for hypothesis H if $x_0$ result from a test procedure with a very low probability or capacity of having uncovered the falsity of H (even if H is incorrect)" (Mayo and Spanos 2009, p. 21). This version of the severity principle is "weak" because it only provides a negative criterion for denying that the result of some test provides evidence for a particular hypothesis. The more general version of the severity principle also specifies conditions under which data generated by a test provide evidence in favor of a hypothesis, namely when the hypothesis has passed a severe test with that data (Mayo 1996; Mayo and Spanos 2006, 2009). But the weak version of the principle is sufficient here due to our concern with identifying cases in which non-epistemic values have inappropriately influenced scientific research.

Consider the following list of "widely used tricks of the trade" for making a newly developed pharmaceutical look more effective or safe than it actually is:

1. Test your drug against a treatment that either does not work or does not work very well.
2. Test your drug against too low a dose of the comparison drug because this will make your drug appear more effective.
3. Test your drug against too high a dose of the comparison drug because this will make your drug appear less toxic.
4. Publish the results of a single multicenter trial many times because this will suggest that multiple studies reached the same conclusion.
5. Publish only that part of a trial that favors your drug and bury the rest of it.
6. Fund many clinical trials but publish only those that make your product look good. (Michaels 2008, pp. 149–50)

All of these tricks of the trade conflict with the severity principle. For instance, the case of a trial that tests a new drug against too low a dose of a comparison claims to provide evidence for the hypothesis that the new drug is more effective than the old one despite the fact that the study had little or no chance of finding this hypothesis mistaken. Moreover, it seems unlikely that there would be any epistemic or practical or ethical reason for not using the normally recommended dosage of the comparison drug (if anything, one would presume that ethical research would require this). Likewise, practices such as over-counting positive results and disregarding contrary evidence, either within a single study or at the level of a meta-analysis of a number of studies, are means of insulating a favored hypothesis from severe tests. All of the "tricks of the trade" listed above, then, are contrary to the values-in-science standard.

However, the values-in-science standard allows non-epistemic values to influence the design and interpretation of scientific research so long as they do not lead to conflicts with epistemic values. This situation could arise if no epistemic value provides a basis for accepting one of several options, or if two or more epistemic values conflict. Thus, the values-in-science standard allows non-epistemic values to influence the design and interpretation of scientific research when the implications of epistemic values for these matters are indeterminate. Consider how the values-in-science standard addresses the objection that an epistemic construal of PP would undermine scientific integrity. A good example of this objection occurs in an article by John Harris and Søren Holm, who suggest than an epistemic PP would recommend exaggerating evidence of hazards to human health and the environment while de-emphasizing contrary evidence (2002, p. 361):

> it is difficult to imagine any justification for an epistemic rule requiring a systematic discounting of evidence pointing in one direction, but not the other.
>
> Such systematic discounting would systematically distort our beliefs about the world, and necessarily over time, lead us to include a large number of false beliefs in our belief system. (Harris and Holm 2002, p. 362)

However, systematic discounting of evidence supporting an undesired conclusion, as illustrated in items 5 and 6 of the "tricks of the trade" list, is prohibited by the severity principle and hence is impermissible according to the values-in-science standard.

Before proceeding, it will be helpful to head off a potential misunderstanding. The value-in-science standard has sometimes been characterized as a "tiebreaker" proposal, according to which non-epistemic values may influence scientific judgments whenever a choice among hypotheses is underdetermined by evidence (Diekmann and Peterson 2013; Elliott and McKaughan 2014). According to the tiebreaker proposal, if evidence does not suffice to indicate which of several hypotheses one should accept, one may select whichever hypothesis is thought to best promote one's political ideals.[2] However, characterizing the values-in-science standard in this way is inaccurate and misleading. It is inaccurate, because there are circumstances in which it would be a very bad idea from an epistemic perspective to accept a single hypothesis when others exist that account for the data equally well. For example, if the accepted hypothesis were to function as the basis for research in a discipline to the exclusion of alternatives, then such acceptance would constitute an unwarranted dogmatism capable of substantially impeding scientific knowledge. The science-in-values standard, therefore, does not grant a blanket dispensation to accept whatever one likes whenever the correct hypothesis is underdetermined. Whether such acceptance is permissible according to the values-in-science standard depends on, among other things, what role the accepted hypothesis will play in ongoing research. The tiebreaker characterization is also misleading, because it suggests that values enter only at the tail end of the scientific process, after value-free data have been compiled. But value judgments can, and often do, arise earlier in the scientific process, for instance, in decisions regarding experimental design or methods for generating an estimate from data (Douglas 2000).

[2] See Intemann (2004) and Howard (2006) for discussion of other philosophers who might be interpreted as advocating such an idea.

### *8.3.2 Uncertainty factors again*

Uncertainty factors are motivated by the value of protecting human health, and are permissible according to the values-in-science standard, so long as the influence of non-epistemic values does not conflict with epistemic values. A critic might argue that uncertainty factors do involve such a conflict because epistemic values would, in a situation of substantial scientific uncertainty, unequivocally demand the suspension of judgment. Thus, one might argue that no reference dose should be proposed until such time as a safe level of exposure for humans can be unambiguously determined. In this section, I explain why this view is mistaken and how uncertainty factors are compatible with the values-in-science standard.

One type of epistemic indeterminacy occurs when countervailing epistemic values are present and no optimal trade-off can be established. In the present case, I suggest that the relevant trade-off is between reliability and speed of methods. The aim of accepting true claims on some topic while avoiding falsehoods inevitably requires a balance between drawing an inference quickly and waiting for further evidence that reduces uncertainties. Very high standards of evidence can postpone the acceptance of truths, but those same high standards may prevent errors. Lower standards of evidence will have the opposite effect. While it is obvious that avoiding error is an epistemic good, it is worth explaining why speed of inference is also epistemically valuable. The speed of a scientific method matters from an epistemic perspective because adopting quicker methods will, at some future time, result in accepting more informative claims. If method A delivers its result in one year while B takes only one week, then adopting B rather than A will result in the acceptance of a more informative claim within a week's time. Moreover, the informational content of claims, not just their truth or falsity, is a crucial epistemic consideration (as noted in section 7.4.1); otherwise, the aim of epistemology could be achieved merely by accepting tautologies. Thus, if two methods A and B are equally reliable but A is quicker, then A is better on purely epistemic grounds. And if method A is slightly more reliable than B but B is quicker, there may be no decisive epistemic answer as to which is preferable. The trade-off between the speed and reliability of scientific methods, therefore, is a trade-off between two epistemic values.

A slower, more cautious approach to balancing the reliability and speed of inference is commonly thought to be adopted in research science. One reason given for this is that an accepted scientific claim is a potential building block for further results, so that one scientific error will have a tendency to generate more errors (Cranor 1993, pp. 25–6; Peterson 2007, pp. 7–8). This "building block" argument is, in effect, the reason given against the naive tiebreaker proposal discussed at the end of the previous section. However, the building block argument does not show that higher standards of evidence are always better for purely epistemic purposes. Even when results are used as premises for further research, the desire to avoid error must still be balanced against the epistemic costs of suspending judgment for too long. Excessively stringent standards of evidence would impede science by giving scientists very few "blocks" of accepted premises to build upon. Furthermore, not all scientific results are used as foundations for further research. Some are used primarily for some practical purpose, such as setting allowable exposure levels to toxic chemicals or predicting climate trends. Claims about reference doses of chemicals are more like scientific endpoints than building blocks for future knowledge. Thus, uncertainty factors arise in a context wherein the epistemic reasons for preferring slower but more reliable methods to quicker but potentially less reliable ones are less applicable than in cases of research science. As a result, that argument fails to show that the use of uncertainty factors in the estimation of reference doses conflicts with epistemic values.

The above considerations are highly relevant to the argument from inductive risk in general, which is often illustrated by cases wherein a pressing non-epistemic value, such as the protection of human health, provides a powerful reason to draw inferences more quickly. For example, Cranor (1993, chapter 4) argues against following conventional, cautious norms of research science when performing risk assessments of potentially toxic chemicals. Useful risk assessment requires not only drawing reasonably accurate inferences about toxic effects, it also demands that those inferences be drawn in a timely manner. As Cranor observes, the result of slow, cautious approaches is widespread public exposure to a large number of potentially toxic chemicals for which no risk assessment has been performed. Thus, he recommends, "The regulatory challenge is to use presently available, expedited, approximation methods that are nearly as 'accurate' as current risk assessment procedures, but ones which are much faster so that a larger universe of substances can be evaluated" (1993, p. 103). Cranor considers two such expedited procedures and compares their results to those produced by slower, more conventional risk assessment approaches, finding that they do not differ too greatly (1993, pp. 137–46). Cranor's argument, therefore, is a good example of how non-epistemic values could influence scientific inferences without compromising epistemic values. From a purely epistemic point of view, the choice between expedited and slower risk assessment methods is a trade-off: quicker inferences versus a somewhat greater chance of error. Cranor's strategy is to argue that the expedited procedures he describes fall into the class of methods that strike a reasonable balance between these two epistemic concerns (1993, p. 142). Thus, from a purely epistemic perspective, neither has a clear advantage over the other, while the expedited procedures are far superior when it comes to reducing the public's exposure to toxic substances, thereby making the expedited procedures the best option overall.

### *8.3.3 Direct versus indirect roles of values*

In this section, I consider an alternative proposal that is founded on a distinction between the direct and indirect roles of values. Although this distinction has been suggested in a variety of ways by a number of philosophers (Heil 1983; Hempel 1965; Kitcher 2001, 2011; Nagel 1961), Douglas (2009) provides the most thorough defense of it, and her version of the position is the focus here. According to Douglas, when values play a direct role, they "act as reasons in themselves to accept a claim, providing a direct motivation for the adoption of a theory" (Douglas 2009, p. 96). By contrast, values playing an indirect role "act to weigh the importance of uncertainty about the claim, helping to decide what should count as sufficient evidence for the claim" (2009, p. 96). Notice that this distinction focuses on how values influence scientists' reasoning (cf. Douglas 2004, p. 460). When values play a direct role, scientists draw a certain conclusion or select a particular methodology *because they want a result that fits their values.* By contrast, when values play an indirect role, they select a methodology or standard of evidence *because they are concerned about harmful consequences of error.* Philip Kitcher expresses a similar idea, referring to "the clear and important distinction between judging that the data are good enough, given the more or less precisely envisaged consequences for human welfare, and accepting a conclusion because it will make you money or advance your favorite political cause" (2011, p. 164). The latter situation would be values acting in a direct role, while the former would be those values acting in an indirect role. One might question whether the distinction between direct and indirect roles of values is drawn with sufficient clarity (see Elliott 2011b). Although I sympathize with such concerns, the objections I raise to Douglas's proposal here do not hinge on this issue.

The central thesis of Douglas's proposal is that while values in an indirect role "can completely saturate science, without threat to the integrity of science" (2009, pp. 96, 115), values in a direct role must be strictly prohibited from infiltrating certain key aspects of the scientific process. Douglas proposes that values in a direct role can legitimately influence which scientific projects to undertake and which to fund and may require prohibiting or interrupting certain types of studies for ethical reasons (2009, pp. 99–101). But values in a direct role are not allowed to impact scientific research in ways beyond these. For example, values in a direct role would not be permitted to guide the choice of study design so as to maximize the chance of a result that supports a favored hypothesis. Similarly, values in a direct role should not be a reason for rejecting, ignoring, or suppressing a study that generates an unwanted result. According to this approach, we have good reason to distrust and discount claims of supposed scientific experts whose statements are directly influenced by values in these ways (see Douglas 2006).

Let us turn, then, to a comparison between Douglas's proposal and the values-in-science standard, beginning with areas of overlap. The severity principle is an important motivation for Douglas's strictures on the direct roles of values:

> One cannot use values to direct the selection of a problem and a formulation of a methodology that in combination predetermines (or substantially restricts) the outcome of a study. Such an approach undermines the core value of science – to produce reliable knowledge – which requires the possibility that the evidence produced could come out against one's favored theory. (2009, p. 100)

The claim that genuine scientific testing "requires the possibility that the evidence produced could come out against one's favored theory" is a formulation of the (weak) severity principle discussed in section 8.3.1. Thus, the severity-compromising direct roles of values that Douglas inveighs against would also be prohibited by the values-in-science standard. The approaches also agree in endorsing the argument from inductive risk and rejecting the value-free ideal (2009, pp. 104–5). Thus, Douglas would presumably regard uncertainty factors as an example of values legitimately operating in an indirect role. Yet *unlike* Douglas's proposal, the values-in-science standard does not give values in an indirect role free rein: it insists that values in this role must not conflict with epistemic values.

More specifically, just like values acting in a direct role, values acting indirectly can conflict with the severity criterion. To illustrate, imagine a chemical corporation that is concerned about uncertainties associated with studies designed to test whether one of its products causes adverse health effects. The corporation judges that it is far worse to mistakenly reject a safe chemical as harmful to human health than it is to make the opposite error. The corporation may make this value judgment because it is concerned to generate profit or because its members sincerely, though perhaps self-servingly, believe that the benefits of new chemicals generally far outweigh the risks. So, the corporation demands a very high standard of evidence for accepting results suggesting adverse effects, and it requires that studies it funds on this topic be designed so as to minimize the chance of falsely concluding that the chemical is harmful. The unsurprising result is a series of very un-severe tests finding no statistically significant adverse effect.[3] How might Douglas respond here? One possibility would be to limit attention to extreme cases in which a completely unattainable standard of evidence for accepting negative results ensures that a favored hypothesis will never be rejected. In such cases, Douglas could say that the difference between the direct and indirect roles of values vanishes. This response is problematic in two ways. First, it is not required that the company's standard of evidence for accepting negative results be unattainable, only that it be substantially higher than that for accepting positive results. For instance, if studies find the drug obviously harmful and ineffective, the company may see no profit in pursuing its development and hence allow publication. Second, the response admits that values in an indirect role are *not* always compatible with scientific integrity, which amounts to granting that there need to be restrictions on the indirect roles of values. A second response is that values in an indirect role must be used in a transparent way, not in a behind-closed-doors manner as in this example (see Douglas 2008). That is sensible but grants that values in an indirect role can be a problem when transparency is difficult to achieve. Examples of such circumstances include highly technical aspects of study design or interpretation of data that people lacking the relevant scientific expertise would have difficulty understanding and cases in which individuals are unaware of their own biases. Finally, Douglas might suggest that values in an indirect role cannot conflict with the severity principle because the severity principle is a rule for assessing how strong the evidence is, while values in an indirect role merely help us decide how strong the evidence needs to be to merit accepting the hypothesis. However, in light of the foregoing discussion we can see that this response presumes, at the very least, that the standards of evidence that are chosen on the basis of values are not too heavily skewed, that the role of values is transparent, and that their effects on severity are clear to

[3] For instance, see vom Saal and Hughes (2005).

all concerned. When these conditions fail, values in an indirect role can generate conflicts with the severity principle.

A second difference between Douglas's approach and the values-in-science standard has to do with application. Applying Douglas's approach requires knowing *how* values figured into scientists' reasoning. Yet values acting in an indirect role could generate the same problematic behaviors as values acting in a direct role (Elliott 2011b, pp. 320–1), as the example from the above paragraph illustrates. Given the difficulties of discerning a person's true reasoning processes, therefore, it may often be impossible to ascertain whether a particularly problematic study design and interpretation of data resulted from values acting directly or indirectly. Indeed, there is evidence that people are often incapable of accurately recognizing how biases, such as those relating to gender, impact their own reasoning (Valian 1998). Thus, people may frequently be unreliable judges of whether values played a direct or indirect role in their own thinking. And the difficulties of discerning precisely how values influenced reasoning are further compounded when one is concerned with groups or organizations rather than single individuals. In contrast, the values-in-science standard does not require discerning the "true" reasoning of individual researchers. What matters from this perspective is the effects those values have, not the psychological mechanisms through which they work on scientists' minds (see Steel and Whyte 2012).

A final difference is that Douglas's proposal treats some epistemic values, such as testability, in the same manner as non-epistemic values, such as the aim to generate profit (Douglas 2009, pp. 106–8). When it comes to indirect roles, both may operate freely, but the direct roles of each in science are restricted in the ways described above. The science-in-values standard, on the other hand, treats testability very differently from non-epistemic values. Testability is a good example of an extrinsic epistemic value, as discussed in section 7.4.1. In contrast, the aim of attaining profit is not an epistemic value at all. Thus, the profit motive is constrained by the science-in-values standard in ways that testability is not. An example will be helpful to illustrate this difference. Consider intelligent design creationism (IDC), which promotes as a scientific hypothesis the idea that an unknown intelligent agent acting by unspecified means is the explanation of complex biological adaptations. Critics of IDC often charge that it is untestable and hence not a scientific hypothesis at all (Pennock 2001). Is this criticism a legitimate invocation of epistemic values in science? Since testability is an epistemic value, the answer to this question from the perspective of the values-in-science standard depends mainly upon the accuracy of the claim (i.e., whether IDC really is untestable) and on philosophical questions about "demarcation" (i.e., how to distinguish science from non-science). If the claim is accurate and if testability is indeed a reasonable necessary condition for a hypothesis to qualify as scientific, then the critique is entirely legitimate. However, from the perspective of Douglas's proposal, the matter differs substantially, since testability seems to be operating in a prohibited direct role in the critique of IDC. To emphasize this point, observe that according to Douglas's proposal, dismissing IDC on the grounds of its lack of testability is no different than dismissing it because it conflicts with one's secular political ideals. In neither case may values operate directly as a reason to reject a claim. Instead, they could only come into play by assessing the consequences of rejecting IDC when true versus accepting it when it is false. The values-in-science standard, by contrast, insists that there is a significant difference between rejecting IDC because it is untestable and rejecting it for political reasons. Rejecting hypotheses from the field of science because they conflict with one's political ideals threatens scientific integrity in ways that dismissing utterly untestable hypotheses does not.

All of the differences surveyed thus far have favored the values-in-science standard. However, one appeal of Douglas's approach is that it appears to avoid relying on the contentious distinction between epistemic and non-epistemic values. But a closer examination of her proposal suggests that this appearance is illusory. Douglas distinguishes values from what she terms "criteria," such as internal consistency and empirical accuracy, that are directly tied to the ultimate aim of science to produce "reliable knowledge" (2009, p. 94). Douglas insists that only epistemic criteria should play a direct role in the evaluation of scientific hypotheses. But then "epistemic criteria" appear to be playing the part of epistemic values. Douglas provides the following reason to think that this is more than a terminological change: "so-called 'epistemic values' are less like values and more like criteria that all theories must succeed in meeting. One must have internally consistent theories; one must have empirically adequate/conforming/predictively competent theories. Without meeting these criteria, one does not have acceptable science" (2009, p. 94).

I do not think this is a reason for denying the label "epistemic value" to consistency and empirical accuracy, which are included among Thomas Kuhn's well-known list of scientific values of theory choice (Kuhn 1977). Values that are fundamentally important for truth-seeking aims of science are still values. Avoiding reliance on the contested distinction between epistemic and non-epistemic values, therefore, cannot be counted among the virtues of Douglas's proposal.

### *8.3.4 Objections from left and right*

This section considers objections to the values-in-science standard from two directions, "left" and "right." Rightward objections charge that the values-in-science standard is too permissive, while leftward objections claim that it is too conservative and that a more extensive role of values should be allowed. I begin with the rightward objections. Perhaps the simplest such objection would be that the values-in-science standard is mistaken because it allows non-epistemic reasons to influence scientific decisions about what hypotheses to accept (cf. Haack 1998, p. 128; Intemann 2004, pp. 1007–8). However, this objection merely reasserts the ideal of value-free science. Since reasons for rejecting the value-free ideal were discussed in Chapter 7, I will not consider this objection further.

A more challenging rightward objection claims that the values-in-science standard is too permissive because "conflict with epistemic values" is an excessively vague concept. The idea that whether a value is epistemic or not may depend on features of context in which it is invoked (as discussed in section 7.4), the objection might continue, further accentuates this difficulty. Thus, Kevin Elliott objects that the account of the distinction between epistemic and non-epistemic values proposed here makes it "very difficult to determine in any particular context which values are in fact epistemic" (Elliott 2013, p. 379). In response, I point out that vagueness is inevitable to some extent in any standard concerning values and science. Indeed, since the value-free ideal relies on a distinction between epistemic and non-epistemic values, it too would be subject to vagaries of this distinction. Furthermore, vagueness in the values-in-science standard can be mitigated in two ways: by (a) focusing on conflicts with robust epistemic values and (b) noting that values can often be expressed in the form of explicit rules. The severity criterion discussed in section 8.3.1 is a good illustration of both points. It is a robust epistemic value, so worries about the context-relative nature of some epistemic values generally do not pertain to it. Moreover, although the severity criterion is certainly not the only epistemic value relevant to the values-in-science standard, it is often what gets violated in examples of "bent" or "politicized" science (McGarrity and Wagner 2008; Michaels 2008). The reasons for this are fairly obvious: if you want to ensure that a scientific study will support a desired hypothesis no matter what, then you need an un-severe test. Finally, severity is not some ineffably vague value but can be expressed in the form of a rule that can be applied to quantitative statistical examples (see Mayo and Spanos 2006).

The objector might reply that this is not an adequate answer because the applications of rules can be vague, for instance, in deciding how un-severe a test needs to be before it conflicts with the severity criterion. However, I think that this reply misconstrues the severity criterion. The severity criterion does not prohibit all tests that fail to attain some specified level of severity. Instead, it says that strength of evidence is proportional to the severity of the test. Thus, a conflict with the severity principle would consist of claiming that a minimally severe test of a hypothesis provides strong support for it. In this way, it would provide a basis for critiquing a study that confidently proclaims to provide strong support for a particular hypothesis despite relying on questionable assumptions that substantially weaken its severity.[4] Similarly, the severity criterion would not be compatible with claiming that a reference dose estimated from animal extrapolation supplemented by uncertainty factors is an "established scientific fact." Reference doses estimated by such means are very uncertain but potentially reasonable guesses that, hopefully, err on the side of protecting public health and might be usefully employed as premises in certain environmental policy decisions.

Let us turn, then, to the leftward objections. According to the first such objection I consider, scientists may have many aims other than discovering the truth, so there is no reason why epistemic values should always trump non-epistemic ones in the design and interpretation of scientific research (Elliott 2013, p. 380; Elliott and McKaughan 2014).[5] In response, I insist that there is a reason for prioritizing epistemic values in the design and interpretation of scientific research, as proposed in the values-in-science standard, namely to maintain scientific integrity. For example, suppose on the contrary that non-epistemic values were allowed to trump epistemic ones, as the objection proposes. Consider, then, a group of scientists doing research for a pharmaceutical company to assess the efficacy of a new drug. Imagine that the scientists carrying out the research decide to utilize some of the "tricks of the trade" listed in section 8.3.1 (e.g., administering too low a dosage of the comparison drug) in order to achieve one of their central aims, namely to get a result to their sponsor's liking and thereby to secure future funding. If non-epistemic values can override epistemic ones in the

[4] See Steel and Whyte (2012) and vom Saal and Hughes (2005) for examples that illustrate this pattern.

[5] Elliott and McKaughan (2014) provide several examples to support this claim. However, all of their examples are instances of the trade-off between slower but more reliable and quicker but less reliable methods discussed in section 7.3.2. Indeed, one of their examples (Cranor 1993) is discussed in section 7.3.2 and in Steel (2010). Consequently, I think that their examples fail to make a case for their thesis.

design and interpretation of scientific research, then why is this behavior problematic?

In response to such concerns, Elliott and Dan McKaughan (2014) suggest that non-epistemic values should be allowed to override epistemic ones only when (a) they operate in a transparent way and (b) they promote the goals of the scientific investigation. However, in my view these two requirements do not come close to effectively addressing the issue. The demand that values promote the goals of the scientific investigation is no remedy for "bent" or "politicized" science. For suppose that the aim of the research is to increase the revenues of a pharmaceutical company or to advance a particular political ideology. Then special interest science of the sort described in the previous paragraph may promote the aims of inquiry. This point can be expressed in more general terms as a dilemma. If the goals of the investigation could be promoted through the promulgation of blatant falsehoods, then values promoting those goals are likely to compromise scientific integrity. But if achieving the goals of the inquiry depends on getting at the truth, then overriding epistemic values will be likely to compromise those goals. In the first case, Elliott and McKaughan's proposal utterly fails to safeguard scientific integrity, and in the second it collapses into the values-in-science standard.

In addition, transparency, while desirable, is insufficient as a philosophical standard and as a practical measure. From the perspective of a philosophical standard, transparency is a means to an end. The hope is that if scientists have to be transparent about how values influence their research, it is less likely that values will influence their research in unacceptable ways. But transparency is useful only provided there is some independent idea of the difference between legitimate and illegitimate influences of values in scientific research. Even if we know the role that values played in a particular case, we can still ask whether that role was appropriate or not. And a normative proposal about the role of values in science is supposed to help answer questions of this sort. Second, as a practical matter the usefulness of transparency is limited for reasons discussed in the previous section. People are often not fully aware of pernicious influences of values in their own reasoning and hence unable to make such influences transparent to others.[6] And transparency is less helpful when non-epistemic values influence highly technical assumptions that would only be understood by people with a high level of domain-specific expertise.

[6] Indeed, Nagel (1961, pp. 489–90) noted this practical limitation of transparency long ago.

The final objection I consider is that the science-in-values standard is too narrowly focused on the design and interpretation of scientific studies and fails to address "big picture" questions about what the overall aims of science should be. Should the promotion of justice, prosperity, and human welfare be basic aims of science? If so, what implications would this entail for which topics get studied (and funded) in science and which do not, and who should make those decisions and how? In response, I agree that the values-in-science standard is not the whole story and that the sorts of questions just raised are important and deserve serious attention from philosophers of science (see Kitcher 2001, 2011; Kourany 2010; Reiss and Kitcher 2009). Indeed, I briefly discuss broader structural issues related to scientific research in the next section and in two case studies in Chapter 9. Nevertheless, the question addressed by the values-in-science standard remains an important part of the picture. If the value-free ideal is rejected, as Chapter 7 argues it should be, then some alternative proposal is needed for how values can influence research without compromising scientific integrity.

## 8.4 Epistemic precaution

In this section, I develop two themes of an epistemic interpretation of PP in relation to the foregoing discussion of uncertainty factors and the values-in-science standard. The first of these two themes concerns the epistemic implications of MPP. As discussed in Chapter 2, MPP recommends against decision-making methods that are susceptible to paralysis by scientific uncertainty. As a consequence, MPP recommends against scientific methods that are excessively slow and cautious in the face of threats to human health and the environment. In addition, MPP is relevant to broader considerations about how scientific research is structured and funded. Actual funding structures for science may fail to provide timely incentives for trustworthy research on crucial topics related to human health and the environment. In such cases, an epistemic MPP supports initiatives to alter these structures, for example, in the form of a requirement for pre-market toxicological testing of chemicals or measures to safeguard against the corruption of scientific research by interested parties. A case study of an epistemic application of PP is discussed in the next chapter. In this section, my chief concern is with a second theme relating to epistemic precaution and the distinction between quantifiable and unquantifiable threats.

Uncertainty factors were chosen as a case study to illustrate the values-in-science standard in part because they are a standard and relatively uncontroversial component of risk assessment. However, the conventionality of uncertainty factors makes them an object of suspicion for some advocates of PP (Quijano 2003, p. 23; Tickner 1999, p. 163). In particular, advocates often insist that PP focuses solely on *hazards*, whose probability and severity cannot be quantified, while cost–benefit analysis and uncertainty factors concern quantifiable *risks* (von Schomberg 2006; Whiteside 2006, pp. 48–9). Thus, it is sometimes claimed that uncertainty factors have no relevance to PP at all.[7] For their part, critics of PP typically do not reject uncertainty factors but instead take them to illustrate that PP is unnecessary and, if interpreted sensibly, no different from current regulatory practices (Marchant 2001a, p. 150; Posner 2004, p. 140; Soule 2004; Sunstein 2001, pp. 103–5). Joel Tickner and David Kriebel, staunch proponents of PP, characterize this objection as follows: "Defenders of the current 'sound science' system argue that there is no need for the precautionary principle or changes in science because uncertainties are addressed through the use of safety factors which toxicologists and risk assessors use when setting so-called 'safe' levels of exposure to chemicals and other agents" (2006, p. 54). Advocates of PP such as Tickner, therefore, often feel obliged to dismiss uncertainty factors as not genuinely precautionary:

> But safety factors are only useful to the extent that the hazardous effects can be accurately observed and their probability quantified. We must also be confident that we are actually studying the most important effect and that we correctly identify the population that will be most affected. Recent research indicates that current safety factors may not be precautionary at all. U.S. Environmental Protection Agency . . . researchers have found that many of the "reference doses" . . . may in fact correspond to disease risks of greater than 1 in 1000 . . . . (Tickner and Kriebel 2006, p. 55)

Tickner and Kriebel are surely correct to claim that uncertainty factors *alone* do not guarantee a truly precautionary approach to environmental policy. That should be no surprise given that uncertainty factors are just one small component of the process. So, the use of uncertainty factors is no reason to think PP is unnecessary.

But advocates of PP sometimes take this point a step further to claim that uncertainty factors are not precautionary at all (Quijano 2003). The reasoning behind such claims is that uncertainty factors are an aspect of a status quo, in which environmental regulations must be justified by

[7] Not all advocates of PP take this position. For example, Weckert and Moor (2006) advocate PP and regard uncertainty factors as an example of it.

quantitative estimates of risks, an approach that often results in scientific uncertainties being used as a reason to stymie attempts to regulate toxic chemicals. This theme is evident in Tickner and Kriebel's statement above, when they distinguish uncertainty factors from PP on the grounds that the former are applicable only when the probability and severity of harms are quantifiable. Other advocates of PP make similar claims. For example, René von Schomberg (2006) distinguishes between *risk* (known adverse impacts and probabilities), *unquantifiable risk* (known impacts but unknown probabilities), *epistemic uncertainty* (serious scientific grounds for suspecting harmful impacts, but their scope, severity, and risk is not quantifiable), and *hypothetical/imaginary risk* (no serious scientific basis for adverse impacts). He claims that PP is applicable only in the middle two of these four situations, unquantifiable risk and epistemic uncertainty (2006, p. 29).

The idea that PP is applicable only in cases wherein probabilities cannot be quantified was discussed and rejected in Chapter 5. I will not review those arguments here, except to note the similarity between the position described in the above paragraph and the decision-theoretic contrast between risk and uncertainty critiqued in that chapter. In this section, my interest is to explore the relationship between an epistemic PP and the distinction between quantifiable risks and unquantifiable hazards. The arguments canvassed in the above paragraph speak of this distinction as if it had an entirely objective existence independent of methodological decisions made by scientists and government agencies charged with protecting human health and the environment. The implicit image is of a division between two sorts of problems existing in nature: those that can be subjected to a quantitative analysis and those that cannot. While I would agree that some objectively present aspects of a problem may make quantification more or less difficult, I also think that it is mistaken, and rather naive, to regard the distinction between quantifiable and unquantifiable risk as some naturally existing category. That is because the distinction between what is quantifiable and what is not depends on human decisions about which methods of quantification are taken to be scientifically legitimate. That in turn leads to important questions about what those methods should be. As discussed in Chapter 7, answering such questions depends in part on value judgments about which errors it is most important to avoid, and PP can be relevant to judgments of this kind, as the case of uncertainty factors illustrates.

Let us consider these points in more carefully. Outside of artificial casino-style examples, there are few cases of entirely unambiguously quantifiable risks (Hansson 2009). If I bet $1,000 that a tossed coin will not turn up heads both times in two tosses, then my risk measured in dollars is −$250. However, in more interesting examples, quantitative values of risks are much more difficult to pin down, since the probabilities may be impossible to know with precision and there may be no straightforward way to convert the harm into a number. These points are well illustrated by environmental risks. Suppose that one wished to provide a quantitative estimate of the risk posed by a chemical that is believed to be carcinogenic. To estimate the probability, one would need to know how likely it is that exposure to the chemical would cause a person to have cancer and how likely it is that a person would be exposed. Evidence concerning the potency of the carcinogenic effect might come from animal experiments and epidemiological studies, while exposure assessments might be estimated from water, air, or soil samples. It is easy to appreciate that many substantial uncertainties would be involved in such estimates. Studies can yield conflicting results, and there are uncertainties associated with each type of study: the vagaries of cross-species extrapolation for animal studies; the possibility of unmeasured confounding causes in epidemiological studies; and the problem of inferring actual physiological uptake from levels of the chemical found in the environment. As a result, quantitative assessments of risks for even well-established hazards, such as arsenic, can be highly contentious and uncertain.

In light of the above, labeling an environmental risk "quantifiable" cannot mean that the probability and severity of the adverse effects can be assigned probabilities in some definitive and uncontroversial way – unless, that is, one wishes to claim that there are practically no quantifiable environmental risks at all. According to von Schomberg, there may be statistical uncertainties associated with quantified risks (2006, p. 29). The thought here appears to be that uncertainties in the estimate of the probability of an adverse effect should be themselves quantifiable by statistical methods, for example, by providing a 95% confidence interval around an estimate. However, the types of uncertainty involved in environmental risk assessments – the comparability of the animal model, conflicting results among studies, whether samples are representative, etc. – are of a substantive nature that could be subjected to quantification only given assumptions that would themselves be highly uncertain. Thus, environmental risk assessments will typically involve not only uncertainties but also "meta-uncertainties," that is, uncertainties about just how great the uncertainty is. One might be tempted to simply assert that practically no environmental risks are quantifiable. But that does not appear to be the position of Tickner, who in the passage cited above states that uncertainty factors are used for quantifiable risks, or of von Schomberg, who states that quantifiable environmental risks should be addressed "with a normal risk management approach" which involves attempting to maintain exposures below safe thresholds or at levels as low as reasonably achievable when safe thresholds do not exist (2006, p. 28).

So, does it mean to say, in a context like that described above, that an environmental risk is "quantifiable"? I suggest that to call a risk quantifiable in such a setting is to express one's acceptance of methods that allow for assigning some approximate numerical value to the risk. In other words, one might argue that, despite the substantial uncertainties, it is nevertheless of practical use to have some means of quantifying the seriousness of environmental risks, and hence methods should be devised for this purpose. I should stress that this proposal is *not* a version of cultural relativism about risk assessment.[8] I think that some methods for assessing risks are better than others on epistemic as well as ethical grounds. The point is simply that which environmental risks are quantifiable is not some external given, but instead is a result of methodological decisions made against a background of scientific, legal, and ethical convictions. Moreover, PP has implications for how such decisions should be made. That point is illustrated by the discussion of uncertainty factors and MPP above and by the discussion of intergenerational impartiality and discounting in Chapter 6. Consequently, the distinction between quantifiable and unquantifiable problems cannot be taken as an established given that exists prior to applications of PP, which in turn is a further argument against the idea that PP is applicable only to unquantifiable risks.

One might object to the claim that PP should influence the choice of methods for quantifying environmental risks on the grounds of the traditional distinction between risk assessment and risk management. According to this picture, social or ethical values can legitimately play a role in *risk management*, the political process of deciding what to do in light of scientific estimates of the risk, but not in *risk assessment*, the scientific process of identifying and quantifying risks. The EU's (2000) *Communication from the Commission on the Precautionary Principle* associates PP with risk management and explicitly denies its relevance to risk assessment. It states: "The precautionary principle, which is essentially used by decision-makers in the management of risk, should not be confused with the element of caution that scientists apply in their assessment of scientific data" (EU 2000, p. 3).

8 See Shrader-Frechette (1991, chapter 3) for an excellent critique of such views.

Since the EU perspective on PP is the background for von Schomberg's discussion (2006, p. 19), one could infer that this is his position as well. However, other advocates of PP strenuously reject the idea that it should be restricted to risk management contexts (Hansen, Carlsen, and Tickner 2007, p. 398; Tickner and Kriebel 2006).

The traditional risk assessment versus risk management distinction is grounded in the ideal of value-free science discussed in Chapter 7. Thus, rejecting the value-free ideal undermines the picture of risk assessment as scientific and value free in contrast to risk management as political and value laden but hopefully informed by science. For if the value-free ideal is rejected, as argued for in Chapter 7, methods for quantifying the probabilities of risks could be influenced by value judgments, such as those embodied in PP. But suppose that one did choose to uphold the value-free ideal, as the EU's stance on PP appears to do. In this case, PP would *still* be relevant to quantifying risks. Recall that risk equals probability of harm multiplied by the severity of the harm. Even if one were to agree that ethical values should play no role in decisions about which methods to use for estimating probabilities, no similar claim is plausible when it comes to quantifying the severity of harms to human health or the environment.

Finally, quantification of risks is important for some of the specific recommendations made on the basis of PP. For example, advocates of PP often support a regulatory approach in which producers or users of a potentially hazardous chemical or activity bear the burden of proof insofar as being required to provide evidence of safety in order to obtain approval for entry into the marketplace (Hansen, Carlsen, and Tickner 2007; Raffensperger and Tickner 1999). This stands in contrast to a system in which evidence of harm is required to justify any restriction of use, and in which producers are not obligated to generate data concerning potential hazards. Yet if a pre-market testing approach is to be practically applicable, it cannot require scientific proof of complete safety, which of course is impossible. Instead, regulations must set some standards for what counts as evidence that a chemical does not pose an "unacceptable risk." Judgments of this kind will inevitably involve some quantification of risks.

## 8.5 Conclusions

This chapter presented and defended the values-in-science standard as a replacement for the value-free ideal as a normative guide for how non-epistemic values may legitimately influence the process by which scientists draw inferences from data and illustrated that proposal with a case study of uncertainty factors. In so doing, it provides a concrete example of how PP can function as an epistemic principle and thereby is a basis for critically examining objections to the use of PP in epistemic contexts (Harris and Holm 2002). This chapter also develops two themes of an epistemic interpretation of PP. One of these themes is an epistemic MPP that recommends against scientific methods and social structures that impede drawing timely inferences in the face of serious threats to human health or the environment. The second theme concerns the distinction between quantifiable and unquantifiable risks. How this distinction is drawn depends to a significant extent on what methods of quantification are judged to be scientifically acceptable. The demise of the ideal of value-free science, therefore, entails that non-epistemic values embedded in PP may influence such decisions, which further undermines the idea that PP is applicable only to cases in which risks are unquantifiable. These epistemic precautionary themes are further illustrated in two case studies discussed in Chapter 9.

CHAPTER 9

# *Concluding case studies*

## 9.1 Recapping the central themes

In this concluding chapter, I recapitulate the central features of the interpretation of PP advanced here and then give three relatively brief case studies, each of which highlights a specific aspect of the proposal. According to my interpretation, PP is encapsulated by three interacting core themes: the meta-precautionary principle (MPP), the "tripod," and proportionality. The first of these, MPP, asserts that scientific uncertainty should not be a reason for inaction in the face of serious environmental threats. This principle is called "meta" because it is not a rule that directly guides environmental policy but instead restricts the sorts of rules that should be used for this purpose. The "tripod" refers to the harm condition, knowledge condition, and recommended precaution involved in any application of PP. Last but not least, proportionality demands that the recommended precaution be calibrated to the level of harm and uncertainty. I explicate proportionality in terms of two subsidiary principles that I call consistency and efficiency. Consistency requires that the precaution not be prohibited by the same version of PP that was used to justify it. Efficiency states that, among those precautions that can be consistently recommended, those with less in the way of negative side effects should be preferred.

These three core themes have been developed and elaborated throughout the book. One distinctive central message of my interpretation is the importance of considering meta-rule, decision-rule, and epistemic aspects of PP in tandem. A meta-rule is significant only insofar as it places constraints on actual decision-rules, while attempts to articulate decision-rule formulations of PP must consider the demand of MPP that scientific uncertainty should not be transformed into a reason for inaction in the face of serious environmental threats. Moreover, an examination of epistemic aspects of PP has significant implications for the scope of the principle as a whole because of the central role of the concept of scientific uncertainty. Thus, one of the main results of Chapters 5 through 8 is that PP should not be restricted to situations involving unquantifiable risks. According to the concept of scientific uncertainty defended in Chapter 5, knowledge of probabilities of outcomes does not always eliminate scientific uncertainty. And Chapter 6 makes the case that PP also has implications for how the interests of future generations should be weighted and hence would be relevant even if outcomes could be predicted with certainty. At a more fundamental level, Chapters 7 and 8 make the case that which risks are quantifiable is not some external given but is instead a result of methodological decisions that may be influenced by PP.

Of course, in addition to the grand themes just sketched, many more specific observations and contributions relating to PP were made throughout the course of the book. Among these were scattered applications of the interpretation proposed here to a variety of cases, including climate change, ground-level ozone, and uncertainty factors. Moreover, concrete examples are useful not only for illustrating more specific points but also for tying together a variety of complex interrelated threads. In the remainder of this concluding chapter, then, I examine three studies chosen to illustrate a distinct aspect of PP as interpreted here. The three case studies discussed are climate change, bovine growth hormones, and the EU's Registration, Evaluation, Authorization and Restriction of Chemicals (REACH) legislation. The case of climate change illustrates the process of selecting the harm and knowledge conditions in an application of PP as well as the difference between PP and cost–benefit analysis. The case of bovine growth hormone is an example of an application of PP wherein the primary harm is inflicted upon agricultural animals rather than humans or "nature" as usually conceived. And REACH illustrates epistemic aspects of PP operating at a structural level. After these three cases, I close with a brief discussion of future lines of research relating to PP.

## 9.2 The precautionary principle and climate change mitigation

Chapter 2 discussed climate change mitigation with the purpose of responding to the objection that applications of PP are either vacuous or incoherent. That discussion made the case that PP supports substantial efforts to mitigate climate change, but did not go into specifics. In this section, then, I examine more closely how PP applies to the case of climate change mitigation.

To answer this question it is necessary to specify the harm and knowledge condition in the version of PP being applied. As noted in section 2.4.1 and the Appendix, the first step in this process is to specify the harm condition.

Once the harm condition is fixed, the knowledge condition is set at the least stringent level that allows a policy to be consistently recommended. The harm condition is, to a large extent, exogenously determined and not dictated a priori by PP. Decisions about the desired level of safety ultimately depend on value judgments that, ideally, would be generated from a deliberative democratic process that is sensitive to concerns of those who would be impacted by the decision. However, PP does place some constraints on the choice of harm condition, two of which are relevant to the present question. First, the harm condition must be chosen so that non-paralyzing and informative versions of PP using that harm condition exist. This constraint rules out excessively sensitive harm conditions that would result from any possible action (e.g., someone is made worse off), as well as harm conditions that would be unlikely to arise from any action (e.g., a complete destruction of the Earth by the end of the week). Second, since PP is incompatible with the application of a pure discount rate, as discussed in Chapter 6, it has implications for how harms should be assessed in decisions regarding long-term problems. With these general observations in hand, let us return to the question of what PP recommends about climate change mitigation.

Article 2 of the United Nations Framework Convention on Climate Change (UNFCCC) is the most widely cited international agreement on the harm that climate change mitigation seeks to avoid:

> The ultimate objective of this Convention and any related legal instruments that the Conference of the Parties may adopt is to achieve, in accordance with the relevant provisions of the Convention, stabilization of greenhouse gas concentrations in the atmosphere at a level that would prevent dangerous anthropogenic interference with the climate system. Such a level should be achieved within a time-frame sufficient to allow ecosystems to adapt naturally to climate change, to ensure that food production is not threatened and to enable economic development to proceed in a sustainable manner. (UN 1992, article 2)

The phrase "prevent dangerous anthropogenic interference with the climate system" is obviously vague. A commonly given benchmark for dangerous interference is a rise in average global temperatures by 2 °C (3.6 °F) over pre-industrial levels. However, concerns have been raised about whether the 2 °C level is sufficient to avoid dangerous interference with the climate (Hansen *et al.* 2008), and a 1.5 °C (2.7 °F) limit has also been seriously considered (UNEP 2010). Nevertheless, I adopt the 2 °C convention here, because this remains the most widely accepted benchmark. The harm condition in this case, then, consists of the damaging effects that would result from an increase in average global temperatures greater than 2 °C above pre-industrial levels. As discussed in section 2.4.2, the consequences of allowing global temperatures to significantly exceed the 2 °C threshold can reasonably be characterized as catastrophic.

Given the harm condition, the next step is to specify the knowledge condition. Discussions of climate change mitigation often evaluate emissions scenarios in terms of a sequence of probability thresholds, including "very likely chance" (>90%), "likely chance" (between 90% and 66%), and "medium chance" (between 50% and 66%). Then questions of the following form are asked. What emissions target is required for a very likely chance of preventing average global temperature from rising more than 2 °C? What emissions target is required for a medium chance of preventing the same outcome? Since the harm condition is defined in terms of dangerous effects of exceeding the 2 °C limit, a likely chance of staying within that limit would correspond to a 34% probability of incurring the harm condition. In this context, then, PP entails that the knowledge condition should be chosen so as to minimize the chance of the harm condition to the extent that is compatible with consistency. To illustrate, consider a version of PP whose knowledge condition is a 34% probability and whose harm condition is defined in relation to the 2 °C increase as described above. Then consistency demands that there be at least one technologically feasible emissions scenario with less than a 34% probability of exceeding the 2 °C increase *and* which does not have a 34% or greater probability of generating catastrophic effects of its own. If such an emissions scenario exists, then PP prefers the 34% knowledge condition to one with a higher risk of incurring the harm condition (e.g., 50%). And if a lower probability threshold exists (e.g., 10%) that also results in a version of PP that consistently recommends some emissions scenario, then that lower threshold is preferred over 34%. In sum, one should apply the version of PP that directs us to adopt a policy with the lowest probability threshold that consistently recommends some mitigation policy. The recommended policy, then, would be one compatible with that version of PP.

To get a sense of what such a policy might look like, consider the sort of mitigation efforts that would be needed for a "likely chance" of avoiding the 2 °C increase. According to a report by the United Nations Environmental Program (UNEP 2010), attaining this probability threshold requires the following:

- a peak in global annual emissions before 2020;
- 2020 global emission levels of around 44 $GtCO_2e$ (range: 39–44 $GtCO_2e$);

- average annual reduction rates of $CO_2$ from energy and industry between 2020 and 2050 of around 3% (range: 2.2 to 3.1%);
- 2050 global emissions that are 50–60% below their 1990 levels;
- in most cases, negative $CO_2$ emissions from energy and industry starting at some point in the second half of the century. (UNEP 2010, p. 12)

Regarding such measures, the UNEP report states: "IAM [Integrated Assessment Model] emission pathways . . . suggest that it is economically and technologically feasible to achieve substantial emission reductions. This implies that it is possible to reach emission levels consistent with a 2°C target" (UNEP 2010, p. 28). Consistency, therefore, does not appear to be a barrier to efforts to achieve these targets. This position is reinforced by the discussion of carbon tax shifts in section 2.4.2. Hence, mitigation targets such as those just listed, and perhaps even stronger measures, would be recommended by PP.

This conclusion is reinforced by analysis due to Frank Ackerman and Elizabeth Stanton (2012; cf. Ackerman and Stanton 2013) of uncertainties relating to estimates of the SCC (social cost of carbon). Ackerman and Stanton's analysis responds to estimates published in 2010 by the US Interagency Working Group on the Social Cost of Carbon (Interagency Working Group 2010). This Interagency Working Group report pegged the SCC at approximately $21 for 2010 and $45 for 2050.[1] This estimate was produced by averaging results from three distinct IAMs known as PAGE, DICE, and FUND,[2] which were the models used by Stern, Nordhaus, and Tol, respectively, to generate the SCC estimates discussed in section 2.3. The Interagency Working Group relies primarily on a discount rate of 3%, although it also reports estimates generated using discount rates of 5% and 2.5% (Interagency Working Group 2010, p. 1).

Ackerman and Stanton perform what amounts to a robustness analysis of the Interagency Working Group's results. Their analysis is based only on the DICE model, which tends to produce intermediate results between PAGE and FUND.[3] Ackerman and Stanton rerun this model while varying three uncertain factors: (1) the damage function for the short term, (2) the damage function for the long term, and (3) the discount rate (Ackerman

[1] The Interagency Working Group published an updated report in 2013 that revised the SCC estimated for 2010 upwards from $21 to $33 (Interagency Working Group 2013). Estimates for the social cost of carbon are associated with years, because the cost of emitting a ton of carbon increases as atmospheric carbon concentrations rise.

[2] These names are acronyms: Policy Analysis of Greenhouse Effect (PAGE), Dynamic Integrated Climate-Economy (DICE), and Framework of Uncertainty, Negotiation and Distribution (FUND).

[3] Ackerman and Munitz claim that the FUND model suffers from a number of defects that make it "arguably inappropriate for setting public policy" (2012a, p. 219; cf. Ackerman and Munitz 2012b; Anthoff and Tol 2012).

and Stanton 2012). For (1), they consider a damage function derived from the work of Michael Hanemann (2008), for (2) they consider a damage function for the long term suggested by Weitzman (2009), and for (3) they consider a discount rate of 1.5% as suggested by Stern (2007). Combining (1) and (2) with the original damage function of the DICE model generates four possibilities: the DICE model with its original damage function for long and short terms, Hanemann's damage function for the short term but the original DICE damage function for the long term, the original DICE damage function for the short term but Weitzman's for the long term, and Hanemann's for the short term and Weitzman's for the long term. Ackerman and Stanton label these four models N-N, H-N, N-W, and H-W, respectively, where N is for Nordhaus, the creator of the DICE model, H is for Hanemann, and W is for Weitzman (Ackerman and Stanton 2012).

The procedure, then, is to run these four models on the same emission scenarios used by the Interagency Working Group with discount rates of 3% and 1.5%. The main result is that if Hanemann's or Weitzman's damage function or the 1.5% discount rate is used, then the SCC by 2050 is at or above – and possibly far above – the maximum estimates of the costs of the most aggressive technologically feasible mitigation policies (2012). Ackerman and Stanton take the significance of this result to be the following: "As long as there is a credible risk that the SCC, or damages from a ton of emissions, could be above the cost of maximum feasible abatement, then it is worth doing everything we can to reduce emissions" (Ackerman and Stanton 2012, p. 20). Let us consider Ackerman and Stanton's analysis from the perspective of PP as interpreted here.

Ackerman and Stanton's procedure follows the pattern of the application of PP described above. They aim to show that there are scientifically plausible scenarios in which catastrophic climate change effects result from business as usual but no such scenarios in which catastrophe is caused by technologically feasible mitigation. In addition, as argued in Chapter 6, PP supports intergenerational impartiality and hence the lower discount rate that Ackerman and Stanton consider. Nevertheless, I do disagree with Ackerman and Stanton on the question of what their analysis entails about the relationship between PP and cost–benefit analysis in this case. They suggest that, so long as there is a serious possibility that the SCC is significantly greater than the maximum costs of mitigation, then "cost-benefit analysis [is] . . . functionally equivalent to a precautionary approach" (2012, p. 20; cf. Ackerman and Stanton 2013, p. 86). However, this does not follow. For even if there is a serious possibility that the costs of climate change will far surpass the maximum costs of mitigation, it may also be possible that the costs of mitigation will exceed their benefits. In this situation, which course of action cost–benefit analysis recommends depends on which scenarios you think are most likely, on your discount rate, and how far off in the future you think catastrophically adverse effects (e.g., a 50% loss of global economic output) would occur if they happened. For instance, a high discount rate combined with a damage function in which costs of climate change rise very gradually with increasing temperatures is a straightforward recipe for generating low SCC estimates. Given scientific uncertainty, then, many distinct specifications of the relevant details may lead to conflicting recommendations. Indeed, Ackerman and Stanton's analysis together with the divergent SCC estimates of Stern, Nordhaus, Tol, and the US Interagency Group all strikingly illustrate this fact. The fundamental difficulty for cost–benefit analysis in these circumstances, therefore, is that it provides very little guidance as to which IAMs, and, consequently, which policy option, to choose. In contrast, PP provides clearer guidance about how to proceed. First, as argued in Chapter 6, it is committed to intergenerational impartiality and hence a lower discount rate. Second, rather than recommending that decisions be founded upon an arbitrary best guess of expected costs and benefits, it directs one to seek out the least stringent knowledge condition that enables a precaution to be consistently recommended. That leads to precisely the sort of question posed by Ackerman and Stanton: do plausible estimates of SCC exist that are substantially greater than the maximum estimates of the costs of aggressive climate change mitigation?

## 9.3 Recombinant bovine growth hormone

Bovine growth hormone, also known as bovine somatotropin (BST), is naturally produced in the pituitary glands of cattle. Recombinant BST, known as rBST, differs from ordinary BST primarily in the manner by which it is produced. Instead of being produced in the pituitary glands of cows, it is produced by genetically modified bacteria that contain the gene for BST. This biotechnological innovation makes it possible to manufacture BST in large quantities and, consequently, to inject it into cows as a means of increasing milk production. Approved by the US FDA in 1992, rBST is also approved for use in Argentina, Brazil, Chile, Colombia, Costa Rica, Ecuador, Egypt, El Salvador, Guatemala, Honduras, Jamaica, Lebanon, Mexico, Pakistan, Panama, Paraguay, Peru, South Africa, South Korea, Uruguay, and Venezuela but is not approved in the EU, Canada, Japan,

Australia, or New Zealand. Sold under the trade name "Posilac" by Eli Lilly, rBST increases the milk production of cows by increasing the quantity of insulin-like growth factor (IGF-1). As an Internet search on "bovine growth hormone" quickly reveals, rBST remains highly controversial and an object of public suspicion. Indeed, due to consumer concerns, several large retailers, including Wal-Mart, Safeway, and Kroger, have adopted a policy of not using milk from rBST-treated cows in their store-brand milk and dairy products. So, what is all the controversy about here, and what implications does PP have for the case? In this section, I compare the rationales of the opposite regulatory decisions on rBST by the US FDA and Health Canada from the perspective of PP.

Two primary concerns have been raised regarding rBST: (1) milk from cows treated with rBST may pose human health risks, for instance, due to containing higher quantities of insulin-like growth factor-1 (IGF-1); (2) adverse effects on the health of cows treated with rBST, such as increased rates of mastitis (a bacterial infection of the udder). I focus on concern (2), because the existence of adverse effects on the health of dairy cows resulting from rBST treatment is generally accepted while the existence of harmful human health impacts is not. Indeed, reviews by the US FDA and Health Canada similarly conclude that the available evidence fails to support major concerns on issue (1). The sole difference is the reaction to a study conducted by Monsanto that found an antibody response in some rats that were orally administered rBST. The US FDA dismisses this result on the grounds that the antibody response may have been due to something other than rBST, while the Health Canada review calls for further research on the topic.[4] Moreover, both the US FDA and Health Canada agree that rBST treatment can cause mastitis. So, the chief difference between the regulatory decisions in this case is that Health Canada sees adverse animal health impacts of rBST as a sufficient reason to deny approval while the US FDA does not.

I begin with a consideration of Health Canada's review of harmful effects of rBST treatment on dairy cows. The Veterinary Medical Association review panel commissioned by Health Canada raised the following concerns on this issue.[5]

[4] This is a good example of the argument from inductive risk, discussed in Chapter 7. The difference between the US FDA and Health Canada does not appear to turn on divergent interpretations of the data (both take it to be inclusive), but on what standard of evidence should be required for dismissing a concern versus calling for additional study.

[5] All references in the following bullets are from the Report of the Canadian Veterinary Medical Association Expert Panel on rBST from the website of Health Canada (http://www.hc-sc.gc.ca/dhp-mps/vet/issues-enjeux/rbst-stbr/rep_cvma-rap_acdv_tc-tm-eng.php).

1. *Mastitis*: An increased rate of clinical mastitis, estimated at 25% on the basis of a meta-analysis of available studies; the panel judged that "health management practices could reduce but not eliminate" this increased risk.
2. *Reproductive effects*: Cows treated with rBST are less likely to become pregnant, although the panel suggested that this effect might be managed by delaying injection with rBST until pregnancies are confirmed.
3. *Lameness*: An increased risk of clinical lameness, estimated at 50% from the meta-analysis; the panel concluded that "Many of the cases of lameness involved joints and dairy producers and veterinarians currently have a limited ability to control or eliminate this increased risk."
4. *Culling*: Cows treated with rBST were more likely to be culled, or killed, a risk that was increased to greater extent for multiparous cows (i.e., lactating cows that had already gone through one or more pregnancies). "When considered along with the increased risk of non-pregnancy, the Panel concluded that the use of rBST would likely reduce the lifespan of dairy cattle."

Considering these issues together, the panel concluded that, "there were a number of legitimate animal welfare concerns associated with the use of rBST," although it stated that further research would be required to decide if these effects are "significant." The decision of Health Canada, then, was presumably that rBST should not be approved until these concerns are adequately addressed.[6]

Let us turn then to the reasoning behind the FDA's decision to approve rBST. Surprisingly, the publically available reasoning given by the US FDA for its decision only discusses potential adverse effects on human health, not health effects on cattle. Indeed, even when discussing mastitis, the US FDA only considers whether increased rates of mastitis in dairy cows pose human health risks. Like Health Canada, the US FDA concludes that risks to humans stemming from antibiotic treatment of mastitis in dairy cows are negligible. Yet animal health is a relevant consideration for the US FDA when it comes to animal drugs, so their reasoning cannot merely be that only adverse effects on humans matter in this case. According to Paul Thompson, the US FDA's reasoning on this issue is as follows:

> It seems clear that rBST causes an increase in milk production. Furthermore, it seems clear that something in the physiology of high producing dairy cattle causes a susceptibility to the so-called production diseases (such as

[6] A 1999 EU-sponsored review of the effects of rBST treatment on dairy cow health makes similar claims about the science, while drawing a stronger conclusion about rBST's adverse impact on animal welfare.

> mastitis) . . . Here we have a case where X causes Y and Y causes Z. Z is production disease, a class of outcomes of clear significance with respect to animal welfare. Y is increased milk production, not necessarily of moral significance and X, of course, is rBST. The defenders of rBST seem to be saying that since Y, a class of events not having moral significance modulates between X and Z, then X is not the cause of Z and should not be held responsible for the moral harm associated with Z. (Thompson 2007, p. 126)

However, this reasoning can be questioned on the grounds that the relevant issue ought to be whether treatment with rBST causes serious adverse effects, not the mechanism or means through which it brings them about. By analogy, imagine a drug whose purpose is to increase red blood cell counts in persons suffering from anemia. Now I presume that increasing red blood cell counts in anemic patients is not a morally objectionable aim in its own right. Nevertheless, if it were discovered that this drug sometimes caused adverse health effects related to polycythemia (i.e., excessive red blood cell counts) in some patients, then that would be legitimate concern in a decision about whether to approve the drug.

Thompson suggests that the US FDA's reasoning can be made "a bit more persuasive" if it is reformulated in the following way:

- Since there are other ways of increasing milk production (such as feed regimens or conventional breeding) that are legal, it would be prejudicial to ban rBST.
- Furthermore, there are ways to control the incidence of disease through careful management and to treat resulting diseases using standard veterinary approaches.
- Therefore, no animal health affects (of regulatory significance) are attributable to rBST. (Thompson 2007, p. 126)

Thompson suggests that while this argument may have some merit from a regulatory perspective, it is questionable on ethical grounds (*ibid.*). However, I believe the reasoning is also flawed from a regulatory point of view, both in general terms and given the specifics of the case.

I begin by considering the latter of these two issues, which concerns the second premise of the above argument. The premise that the adverse effects of rBST treatment can be effectively managed by existing veterinary and animal husbandry practices is directly called into question by the Health Canada report on veterinary issues summarized above. The point that some adverse effects of rBST treatment might not be effectively manageable arises not only with respect to mastitis but even more so for lameness. In addition, the increased rate of culling produced by rBST treatment casts doubt on the manageability of its adverse effects.

Turn, then, to the general problem with the argument, namely that the first premise could be given as a reason against regulating almost anything because the sort of situation it describes is ubiquitous. In all but the simplest of circumstances, causes produce effects through intermediaries (e.g., X causes Y which in turn causes Z), and those intermediates themselves normally have multiple causes. Moreover, it is almost never the case that all the factors that affect intermediate causes are subject to regulation. Indeed, causes of intermediates may include things that it would not be sensible to regulate, for instance, because doing so would entail extensive interference with civil liberties. Suppose, then, that it were judged "prejudicial" to regulate an entity X, which causes the adverse effect Z through the intermediate cause Y, whenever there are other causes of Y that are unregulated. The consequence of such a rule would be to prevent regulation in almost all cases. For to block regulation, it would only be necessary to find some unregulated cause that also impacts the intermediary Y. To illustrate these points, consider the hypothetical example described above of a drug that promotes red blood cell production in anemic patients. Many factors – such as genes, exercise, oxygen levels in the ambient air, and diet – can influence a person's red blood cell count. Moreover, not all of these factors are subject to regulation, and some, such as genes and exercise, would be difficult to regulate without compromising civil liberties (e.g., eugenic policies or forcing people to do just the right amount of exercise).

In sum, it would be very unwise to insist that an entity that causes an adverse effect cannot be regulated until all other factors that promote the same outcome have also been regulated. Such an approach would create an absurd situation in which nothing with a specific harmful impact can be regulated until everything is. Why pick on tobacco, for example, when so many other things can cause cancer, including many that aren't regulated – like sunlight?[7] Not only would such an approach hamstring regulation aimed at protecting health and the environment, it is also unreasonable considered on its own terms. If a drug, say, has harmful side effects, then that may be legitimate reason for deciding not to approve it even if there are other unregulated causes that have similar effects. For questions about the regulation of those other causes are separate issues. Perhaps they should be regulated, too. Or perhaps there is some relevant difference that justifies regulating in one case but not the other. But judgments on these matters

[7] In fact, this style of argument by deflection of blame is a familiar tactic in the "tobacco strategy" arsenal (see Oreskes and Conway 2010).

would require separate processes that carefully examined the details of each case.

What, then, does PP entail about rBST? In my view, the answer to this question depends in large measure on how much weight is accorded to the health of agricultural animals, a topic that the interpretation of PP presented in this book has not addressed. However, if one supposes that the health of agricultural animals should be a serious consideration, then Health Canada's decision not to approve rBST is arguably justifiable by PP. I say "arguably" because a full examination of the case would demand a greater level of detail than is possible here. But the outlines of the reasoning based on PP are easy to sketch. Since there seems to be fairly compelling evidence of harmful health effects of rBST upon dairy cows, the version of PP to applied to the case can utilize a strong knowledge condition. As discussed in section 2.4.1, a stronger knowledge condition makes consistency easier to satisfy.

In this case, consistency requires that there not be strong evidence that Health Canada's decision against approving rBST would produce effects as bad or worse than the harmful impacts on animal health. Assuming that health impacts on dairy cows are a serious consideration, there appears to be a solid argument that consistency is satisfied in this case. There is no urgent need to increase milk production in Canada, and the economic benefits of rBST are in any event questionable. Empirical research on the economic advantages of using rBST have found mixed impacts on profitability and high rates of "disadoption," that is, dairy farms adopting rBST and later deciding to discontinue its use (Barham *et al.* 2004; Foltz and Chang 2002; McBride, Short, and El-Osta 2004). These data should not be surprising given the adverse health effects on cattle of rBST treatment described above. Similarly, that Wal-Mart, the king of low-cost retailers, would voluntarily forego milk from rBST-treated cows in its store-brand dairy products suggests that price reductions generated by rBST are rather small. But might there be environmental benefits of rBST that outweigh its harmful effects on dairy cows? One study claims that rBST has positive environmental impacts due to making dairy production more efficient and therefore generating less manure and GHGs per unit of milk (Capper *et al.* 2008).[8] However, this study is based not upon empirical data but instead on a model that assumes that the only effect of rBST is to increase milk output of cows. Consequently, it does not consider that cows treated with rBST

[8] It is of interest to note that one of the authors of this paper reported the following conflict of interest: "a full-time employee of Monsanto, holding the position of Technical Project Manager for POSILAC rbST" (Capper *et al.* 2008, p. 9668).

would eat more, thus consuming more resources and possibly producing greater quantities of manure and methane per cow than those not treated with rBST. The study also ignores adverse health effects of rBST treatment, which would be expected to reduce gains in efficiency and increase culling. The issue of culling is relevant because the claimed environmental benefits of rBST treatment are said to follow from reducing the overall herd size needed to produce a given quantity of milk (Capper *et al.* 2008, p. 9669). Yet a higher cull rate entails that dairy cows would need to be replaced more quickly, which would require raising more heifers and having more lactating cows on standby.

While Health Canada's rationale for not approving rBST is compatible with PP, the same certainly cannot be said for the reasoning attributed to the US FDA by Thompson. In an application of PP, the severity of the harm and the strength of the knowledge condition associated with it matter, but the existence of other unregulated factors that may also cause similar effects is irrelevant. The central question is whether the regulation in question is supported by a version of PP that can be consistently applied in the case. Whether PP might also be used to argue against something else that could lead to similar harms is a separate matter. The reasoning described by Thompson, moreover, is a nice example of a rationalization that can generate "late lessons from early warnings" such as those discussed in Chapter 4. As such, it illustrates the sort of thinking PP is designed to avoid.

Nevertheless, it would be an exaggeration to assert that PP unequivocally comes down in favor of a ban on rBST. The reason, as noted above, is that the case turns in part on how much weight is given to the health and welfare of agricultural animals in regulatory decisions. That is a complex ethical question that PP does not address. Indeed, it is a question that highlights a lacuna in many formulations of PP, which are often framed in terms of threats to human health and the environment. While the welfare of wild animals is encompassed under the rubric of "the environment," the health of agricultural animals does not obviously fall into this category. More generally, the rBST case illustrates that environmental policy issues may depend in part on ethical questions that PP itself does not answer. Similarly, as noted in section 4.4.2, considerations of justice are often relevant to applications of PP, yet PP is not a theory of environmental justice. In my view, this feature of PP is unavoidable given that ethical conundrums are potentially endless. No theory or decision approach could ever hope to adequately address them all. It is also a reflection of the limited function PP is designed to perform, namely to provide a reasonable framework for decision making that is not susceptible to paralysis by scientific uncertainty.

## 9.4 REACH

Registration, Evaluation, Authorization, and restriction of CHemicals (REACH) is a body of legislation that was approved by the European Parliament in 2006 and which went into force on June 1, 2007. REACH requires any chemical manufacture or import into the European Union in quantities of 1 ton or more annually to be registered with the newly created European Chemicals Agency (ECHA) by companies both upstream and downstream in the supply chain. The companies are also required to provide information regarding the chemical, where the specific information demanded varies with the tonnage of the chemical and its classification (see Rudén and Hansen 2010). With one exception to be noted below, REACH treats pre-existing and newly created chemicals similarly, and thus does not allow a "grandfather" exemption for chemicals already in use. After a chemical is registered and the required information submitted, ECHA carries out an evaluation of potential harmful effects of the chemicals. As a result of this evaluation, the chemical may be reclassified, for instance, as persistent, bioaccumulative, and toxic (PBT) or carcinogenic. Companies wishing to use chemicals whose classification puts them into the category of "substances of very high concern" (SVHC) – a category that includes the two classifications just mentioned – must apply for authorization to do so. To receive authorization for the use of a SVHC, a company would have to show that its use fulfills a substantial socioeconomic purpose that could not be achieved by a feasible safer alternative and that the risks posed by the chemical could be effectively managed. Finally, restrictions can be placed on the use of certain SVHCs, including but not limited to bans on particular chemicals. The initial implementation of REACH focused on registering the large number of chemicals in use, including the large number that had been "grandfathered" in under previous chemicals legislation in the EU. As registration dossiers have come in, ECHA has also begun the evaluation process, which has resulted in changes in classifications, usually in the direction of greater stringency (EU 2013, p. 4). In coming years, it is expected that the authorization and restriction components of REACH will be more fully implemented.

It may be helpful to contrast REACH with the United States' Toxic Substances Control Act (TSCA), passed by Congress in 1976 (see US General Accounting Office 2007). Under TSCA, the US EPA generally bears the burden of providing evidence of risk before any rule concerning a chemical can be issued, including rules requiring that testing be performed to assess the chemical's toxicity. Under REACH, by contrast, manufacturers and importers of chemicals are required to provide evidence that the chemical does not pose an unreasonable risk prior to being granted entry to the EU market. There are a number of other important differences between TSCA and REACH. For instance, under TSCA it is more difficult for the US EPA to regulate chemicals already in use at the time of the Act's passage, which constitute the vast majority, than newer chemicals. In contrast, REACH generally imposes the same testing requirements on old and new chemicals alike.

The differences between REACH and TSCA can be conceptualized in terms of a distinction, noted in section 3.4.1, between actively and passively adaptive policies. Adaptive policies are designed to respond to future "signposts," in the form of new information or technological innovations. The policy leaves some future courses of action open, so that the path taken depends on future events that are difficult to foresee at present. An actively adaptive policy also takes steps to make signposts arise more rapidly rather than passively waiting for them to appear. Moreover, there is a straightforward argument that a pre-market approach to chemicals regulation, such as REACH, is much more actively adaptive than a post-market approach, such as TSCA. In a post-market regulatory system, chemicals may be used for commercial purposes without safety testing and are only withdrawn or regulated if evidence of risk arises. In a pre-market system, by contrast, market entry of new chemicals, and continued commercial use of already existing chemicals, is conditional on safety testing being carried out and results being submitted to the regulatory agency for evaluation. Both of these regulatory approaches are adaptive in that they are designed to respond to the emergence of new evidence indicating that a chemical has toxic effects. And, in principle, a post-market regulatory approach could be actively adaptive, since the agency could be empowered to take steps to generate research concerning the toxicity of chemicals. However, in practice, post-market systems are often passive, because (1) the agency does not have the resources to carry out experimental tests of chemicals itself and must ask companies to do so and (2) the post-market system surrenders the most effective incentive the agency could have for getting companies to perform these tests in a timely manner. As a result, a post-market regulatory approach institutionalizes a perverse "ignorance is bliss" incentive: since the use of a chemical can be restricted only if scientific research documents its harmful effects, companies involved in the use and manufacture of chemicals have a powerful incentive not to carry out such research (Case 2005; McGarrity and Wagner 2008, pp. 104–7). And since the agency cannot force companies to produce the data, prolonged delays are a predictable result. As a result of this situation, critics often deride TSCA as the "toxic substances conversation act" (Lohmann, Stapleton, and Hites 2013, p. 8995). In contrast, a pre-market system provides firms with the strongest possible incentive to perform toxicity testing of chemicals in accordance with specified deadlines (i.e., testing is required for commercial use).

Since adaptive management is often associated with a precautionary approach to environmental policy (see section 3.4), the above discussion suggests a connection between REACH and PP. In addition, PP is often associated with the idea that the burden of providing evidence of safety should be borne by those who propose to use some new substance or processes. It should not be surprising, then, that REACH explicitly cites PP as an underlying rationale (EU 2006). However, some questions have been raised about the extent to which REACH and PP truly coincide (Hansen, Carlsen, and Tickner 2007; Rudén and Hansen 2010). For example, one significant concern is that, despite the "no data, no market" slogan, REACH does not ensure that toxicity data will be generated for all chemicals surpassing the 1 ton threshold. For example, no toxicity testing is required for non-prioritized phase-in[9] chemicals falling in the 1 > 10 tonnage band (Rudén and Hansen 2010, p. 6). Depending on the criteria for classifying phase-in substances as "non-prioritized," this could become a loophole through which potentially hazardous chemicals evade scrutiny. But my focus here is not with the specifics of the REACH legislation but with examining the connection between a pre-market system of chemicals regulation and an epistemic interpretation of PP.

In particular, I claim that implementing a pre-market testing approach to chemicals regulation, such as REACH, is an application of the argument from inductive risk discussed in Chapter 7. The argument from inductive risk asserts that ethical value judgments about the severity of errors should influence decisions about what should count as sufficient evidence for accepting a claim. In effect, a pre-market approach makes toxicity the default assumption for chemicals. The obligation, therefore, is on the users and producers of the chemical to provide evidence that this default assumption is false and that the chemical can be safely used. But what one treats as the default assumption reflects a judgment about the relative costs of opposite possible errors. Thus, REACH is motivated by the judgment

[9] A "phase-in" chemical is one that was already in commercial use at the time of the implementation of REACH.

that it is normally better to delay the commercialization of a safe chemical than to belatedly discover that a chemical already in widespread use is seriously harmful.

The case illustrates a feature of practical applications of epistemic PP that it shares with the example of uncertainty factors discussed in Chapter 7, namely that it operates at the level of an established institutional practice rather than as a matter of individual prerogative. In fact, REACH operates at a broader institutional level than uncertainty factors. While uncertainty factors are a widely recognized practice aimed at erring on the side of caution when producing estimates required within a particular regulatory framework, REACH restructures that institutional framework itself. The observation that epistemic applications of PP can operate at a structural rather than individual level is significant in several ways. To begin with, it is the opposite of the impression one might acquire from reading philosophical discussions of the argument from inductive risk. In such discussions, the imperative to consider the ethical costs of error in decisions about what should count as sufficient evidence is frequently discussed as an obligation or virtue of *individual* scientists (see Douglas 2009, chapter 4; Rudner 1953). However, there are a number of advantages to making such value judgments at a structural level when it is feasible to do so. One is that structural problems often require structural solutions (Anderson 2012). Virtue on the part of individual scientists is unlikely to overcome the harmful effects of a system with entrenched structural incentives not to produce certain types of important knowledge. Additional advantages include greater transparency and easing the burden upon scientists to make value judgments. Decisions about reforms of structural features of a regulatory system can be openly debated in democratic societies, and input from stakeholders and experts can be elicited in the process. Moreover, when decisions about which hypothesis to treat as a default or about the use of uncertainty factors are treated as group or institutional decisions, the onus on individual scientists to make such decisions is reduced. That is helpful if such value decisions involve expertise on topics that scientists may lack.

Nevertheless, I agree with Douglas that it would be impossible and not even desirable to completely remove value judgments relating to standards of evidence from individual scientists or research teams. Instead, the suggestion here is that value judgments involved in scientific research can arise at several distinct levels. Consider these three:

1. A structural framework of a system for generating and applying knowledge (e.g., pre- versus post-market approaches to chemical regulation).

2. General methods or standards of evidence adopted within such a framework (e.g., the use of uncertainty factors in generating estimates of acceptable exposure levels of chemicals).
3. Case-specific methodological decisions (e.g., choice of statistical model for analyzing a data set).

These three levels are intended to be illustrative rather than exhaustive. Moreover, the levels are closely intertwined rather than sharply distinct. For instance, REACH relies on a variety of classifications of substances, which determine, among other things, whether authorization is required for the commercial use of a substance. Such classifications, then, would be features of the general framework of the system (i.e., level 1). But a number of annexes to the REACH legislation specify tests and evidence deemed sufficient for classifying a given substance into one or another category (i.e., level 2). Nevertheless, the three levels distinguished above are helpful for making an important point about epistemic applications of PP. To the extent that PP is regarded as a guide for policy decisions, epistemic applications of PP will tend to be at levels 1 and 2 rather than 3. Individual scientists or research teams might choose of their own accord to apply an epistemic version of PP. But such decisions, while perhaps laudable, would not constitute an application of PP at the level of environmental policy. However, this not to say that such epistemic applications of PP are unimportant. In fact, one advantage of such smaller scale epistemic applications of PP is that they do not require difficult-to-enact changes to deeply entrenched features of existing systems, such as existing frameworks for chemical regulation.

Considering epistemic applications of PP from a structural angle is also significant because it leads to a basic rethinking of the traditional risk assessment versus risk management distinction. It draws attention to the fact that risk assessments are inevitably carried out within a regulatory system wherein substantive value judgments have already shaped epistemic processes. Hence, even if individual scientists working within such a system could somehow operate in a purely value-free manner, the knowledge produced by the system would not be pure and unsullied by value judgments. This observation undermines the traditional risk assessment versus risk management distinction at a fundamental level. Indeed, one concern raised about REACH is that it is linked to an interpretation of PP that retains the traditional risk assessment versus risk management distinction (Hansen, Carlsen, and Tickner 2007, p. 398).

In sum, REACH illustrates that epistemic applications of PP may occur across a range of levels, from individual research projects to fundamental reforms of institutions intended to protect human health and the environment. Appreciating this fact, I believe, is crucial for an accurate understanding of epistemic applications of PP. While such applications may concern the design of a particular scientific study, they often will target structural features of systems through which scientific data are generated and brought to bear on decisions related to public health or the environment.

## 9.5 Future paths for precaution

Debates about PP raise fundamental philosophical questions about uncertainty, rational decision making, scientific evidence, and their mutual interconnections. This book suggests a picture of how these strands can be brought together in a coherent interpretation. But like any picture, the one sketched here is unavoidably incomplete and partial. Many topics related to precaution await further exploration. Among these, I mention just a few. How does the interpretation proposed here connect to public participation, which is sometimes cited as an important plank of PP? What implications does an epistemic PP have for the problem of special interest science, wherein scientific results are distorted when funders have a stake in which conclusions are reached? How should pros and cons be balanced in decision problems in which incommensurability is a serious concern? What is the relationship between PP and sustainability? No doubt other questions have occurred to readers along the way. But I hope that these are sufficient to give a sense of the branching paths forward.

# *Appendix*

## 1 Formalizing the precautionary principle

In this section of the Appendix, I explain how the interpretation of PP I propose can be formalized. This formalization enables some aspects of the interpretation to be expressed more clearly and is shown here to lead to several fundamental results that would be difficult to work through in an informal discussion.

### *1.1 The basic set-up*

Formally, a version of PP is a triple $V = \langle S, H, A \rangle$, where S is a set of scenarios that map actions onto outcomes, H is a set of outcomes that fall below a specified threshold of acceptability, and A is a set of possible actions. Informally, S corresponds to the knowledge condition, H is the harm condition, and the actions in A are recommended precautions. Let $S_\Omega$ be the set of all possible scenarios, $H_\Omega$ the set of all possible outcomes, and $A_\Omega$ the set of all possible actions.[1] Scenarios, then, are functions from actions in $A_\Omega$ to outcomes in $H_\Omega$. I assume that scenarios are mutually exclusive: that is, for any two scenarios at most, one can be true. Likewise, actions are assumed to be mutually exclusive in the sense that no more than one can be enacted.

Versions of PP may differ with respect to the stringency of the knowledge condition, the severity of the harm condition, and the strictness of the recommended precautions. The relative stringency of knowledge conditions can defined as follows.

1. Let $V = \langle S, H, A \rangle$ and $V^* = \langle S^*, H^*, A^* \rangle$ be two versions of PP. Then the knowledge condition of $V$ is *more stringent* than that of $V^*$ if $S \subset S^*$, *less stringent* if $S^* \subset S$, and *equally stringent* if $S = S^*$.

[1] To ensure that these sets exist, I will assume for the present purposes that their cardinality is countable. This assumption could easily be relaxed, but it poses no difficulties for the ensuing discussion.

Thus, less stringent knowledge conditions include a broader range of possible scenarios, while more stringent ones include a more restricted range. Clearly, $S_\Omega$ is the least stringent knowledge condition. Although analogous definitions for the relative severity of harm conditions and relative strictness of recommended precautions could be given, such a concepts are less useful and hence not explicitly defined here.

The notion of greater and lesser stringency is more useful when knowledge conditions are comparable in the following sense:

2. Let $V = \langle S, H, A \rangle$ and $V^* = \langle S^*, H^*, A^* \rangle$ be two versions of PP. Then *the knowledge conditions of V and V* are comparable* if and only if S and S* are equally stringent or one is more stringent than the other.

The comparability of knowledge conditions greatly simplifies applications of PP. Thus, it is desirable that distinct knowledge conditions be defined so as to be comparable in applications of PP. It is also natural that the number of the qualitative knowledge conditions be finite. One simple example of how to achieve a finite comparable ranking of knowledge conditions would be to use a sequence of probability thresholds, as is in fact done in assessment reports of the IPCC. However, it is not necessary that thresholds be associated with quantitative probabilities. Indeed, these should be taken with a grain of salt in the case of the IPCC. All that matters is that the knowledge conditions $S_1, S_2, S_3, \ldots, S_\Omega$ can be arranged into a sequence of monotonically increasing sets (e.g., $S_1 \subset S_2 \subset S_3 \ldots \subset S_\Omega$).

Since many different choices of S, H, and A are possible, many different versions of PP can be formulated. Thus, some further discussion of how versions of PP should be selected and applied is needed. Consistency, which was introduced in section 2.4.1, is a first step in this direction.

### 1.2 Consistency

Informally, consistency requires that the version of PP used to justify a precaution not also recommend against that precaution. Definitions 3 through 5 lead to a more precise statement of consistency:

3. For a version of PP $V = \langle S, H, A \rangle$, the set of *dismal scenarios*, $S_d$, is defined as follows: $s \in S_d$ if and only if $s \in S$ and, for every $a \in A_\Omega$, $s$ leads to an outcome in H.

A dismal scenario, then, is a member of S in which the harm condition results no matter what action is chosen.

4. For any action $a$, $V = \langle S, H, A \rangle$ *recommends against a* if and only if there is a scenario in $S - S_d$ wherein $a$ leads to an outcome in H.

For example, suppose that the knowledge condition in S is characterized as a scientifically plausible mechanism, while the harm condition is catastrophe. Then the version of PP recommends against any action *a* if and only if there is a scientifically plausible mechanism whereby *a* leads to catastrophe, *excluding* those scenarios wherein catastrophe results no matter what action is taken. The exclusion of dismal scenarios reflects the idea, discussed in section 2.4.1 and 3.3.3, that recommending against an action on the grounds that it would lead to harm in a particular scenario is sensible only if some alternative action exists that would avoid that harm in that scenario.[2]

5. For a version of PP $V = \langle S, H, A \rangle$, let the *consistent recommendation set*, denoted by $R(V)$, be the set containing all and only those actions $a$ such that $a \in A$ and $V$ does not recommend against $a$.

I will say that a version of PP $V$ can consistently recommend an action $a$ just in case $a$ is in $R(V)$. Given the above definitions, consistency can be stated as follows:

> *Consistency*: A version of PP $V = \langle S, H, A \rangle$ can be used to justify an action $a$ only if $a \in R(V)$.

The above is equivalent to the formulation of consistency given in section 2.4.1, which stated that a version of PP used to justify a precaution must not also recommend against that precaution. Note that consistency states a necessary condition for justifying an action by means of PP, not a sufficient one.

Consistency is logically stronger than the requirement that versions of PP not be paralyzing (see section 2.5.1). Given the above definitions, paralyzing versions of PP can be characterized as follows:

6. A version of PP $V = \langle S, H, A \rangle$ is *paralyzing* if and only if $R(V)$ is empty.

Paralyzing versions of PP can arise in a number of ways. Extremely lax knowledge conditions can give rise to paralyzing versions of PP, as illustrated by Manson's "wild story" of climate change mitigation leading to nuclear holocaust (see section 2.4.2). However, laxity of the knowledge condition is not the only route to a paralyzing version of PP, which is why a *de minimis* condition fails to answer the incoherence objection (see section 2.5.1). A version of PP might be paralyzing because it includes only extremely drastic and disruptive precautions in the set A, a point illustrated by version 4 of PP given in section 2.5.1.

Obviously, consistency entails that no paralyzing version of PP can be used to justify a precaution. But even if $V$ is not paralyzing, it may be that A

[2] Munthe (2011) incorporates a similar idea in his account of PP, by insisting that a risk has been irresponsibly imposed only if there is an alternative less risky action.

is not a subset of $R(V)$, that is, it may be that $V$ can consistently recommend some of the actions included in A but not others. For this reason, consistency requires not only that the version of PP be non-paralyzing but also that it does not recommend against the specific action being proposed. Thus, consistency is logically stronger than a prohibition on paralyzing versions of PP, and this additional strength is relevant to addressing the incoherence horn of the dilemma objection, as noted in section 2.5.1.

### *1.3 The uniqueness theorem*

Consistency imposes a necessary, not sufficient, condition on the justifiability of an action by PP, since it is possible for consistency to be satisfied with respect to several distinct actions. This may be because a single version of PP consistently recommends more than one action or because more than one version of PP consistently recommends something. In the former case, a decision about which of the consistently recommendable actions to adopt would rest on considerations of efficiency. I am concerned with the latter case here. Conflict in this case arises when $V = \langle S, H, A\rangle$ and $V^* = \langle S^*, H^*, A^*\rangle$ exist such that $V$ recommends against some $a$ in $R(V^*)$. For example, suppose $S^*$ is more stringent than S, but $H = H^*$. Then there may be a (non-dismal) scenario in S through which $a$ leads to an outcome in H but no such scenario in $S^*$. Such examples illustrate the need for some criteria for selecting among non-paralyzing versions of PP that generate conflicting recommendations.

Selecting a harm condition is the first step in identifying an appropriate version of PP to apply in a particular case. I view the harm condition as deriving from deliberations concerning a desired level of protection rather than dictated by abstract principles inherent in PP itself. The choice of harm condition is fundamentally a value judgment and should, whenever possible, be made in consultation with those who would bear the brunt of the potential harm. The harm condition is consequently determined on a case-specific basis rather than fixed across the board. It would be a mistake, for instance, to assume that the harm condition must always be "catastrophe." Consistency does place a formal constraint on the choice of harm condition: the harm condition H must be chosen such that there exists a non-paralyzing version of PP $V = \langle S, H, A\rangle$. However, this requirement is less demanding than it might seem, since a non-paralyzing version of PP can normally be found for a given harm condition by making the knowledge condition sufficiently stringent.

But the prohibition on paralyzing versions of PP is not the only constraint. The flipside of a paralyzing version of PP is an uninformative one, that is, a version of PP that recommends against nothing. This idea can be defined more exactly as follows:

7. A version of PP $V = \langle S, H, A \rangle$ is *uninformative* if and only if $R(V) = A_\Omega$.

In other words, the recommendation set of an uninformative version of PP is identical to the full set of alternative policy options. Whereas extremely lax knowledge conditions could generate paralyzing versions of PP, uninformative versions can result in the opposite manner from excessively stringent knowledge conditions or overly severe harm conditions. To illustrate, consider a version of PP according to which precaution should be taken only if it is absolutely certain that the activity will result in the complete extinction of human life within the next 25 years. Such a version of PP would recommend against no action whatever (not even nuclear war, which might leave some survivors!) and hence is utterly uninformative. Uninformative versions of PP can also result from overly sensitive harm conditions that are sure to result no matter what action is chosen, such as that someone is made worse off. In such cases, every possible scenario is a dismal scenario, $S - S_d$ is empty, and as a result the version of PP trivially recommends against no action (see definition 4 above). Since uninformative versions of PP are useless for guiding decisions, an adequate version of PP should be informative.

The demand that versions of PP be neither paralyzing nor uninformative provides a baseline for adequacy. But two further requirements are necessary for a version of PP to be fully adequate. The first is that a version of PP $V = \langle S, H, A \rangle$ should not arbitrarily exclude actions from A that are admissible given its knowledge and harm conditions. I express this constraint by means of the concept of a complete version of PP:

8. A version of PP $V = \langle S, H, A \rangle$ is *complete* if and only if $R(V^*) = R(V)$, where $V^* = \langle S, H, A_\Omega \rangle$.

Thus, $V = \langle S, H, A \rangle$ is complete just in case it includes all actions in A that it could consistently recommend, so replacing A with the set of all possible actions $A_\Omega$ would make no difference to $R(V)$. I take it as a reasonable demand that fully adequate versions of PP should be complete, since otherwise they arbitrarily exclude some actions. Munthe (2011), for instance, proposes that the avoidance of arbitrary distinctions is a criterion for adequate versions of PP. If there are reasons for not choosing certain members of $A_\Omega$ that $V$ can consistently recommend, those reasons are not given by $V$ itself but stem from other considerations (e.g., efficiency). The definition of a fully adequate version of PP, then, is given below. The fourth and final criterion of a fully adequate version of PP is included as condition iii.

*Adequate versions of PP*: Let $H_p$ be the harm condition selected to provide the desired level of protection. Then a version of PP $V = \langle S, H_p, A\rangle$ is *adequate* if and only if it satisfies the first condition below and *fully adequate* if and only if it satisfies all three.

i. $\varnothing \subset R(V) \subset A_\Omega$ (i.e., $V$ is neither paralyzing nor uninformative),
ii. $V$ is complete; and
iii. there is no non-paralyzing and complete version of PP $V^* = \langle S^*, H_p, A^*\rangle$, such that $R(V^*) \subset R(V)$.

The rationales for the first two conditions in this definition were discussed above. Requirement iii says, in effect, that a fully adequate version of PP cannot be made more cautious by modifying the knowledge condition. Thus, it cannot be that some actions allowed by $V$ would be recommended against by a non-paralyzing version of PP $V^*$ that utilizes a less stringent knowledge condition. This aspect of requirement iii is motivated by MPP: unnecessarily stringent knowledge conditions should not be used as reasons against taking precautions in the face of environmental threats.

From the above definition, it can be shown that, given a harm condition that specifies the desired level of protection and that knowledge conditions are comparable, every fully adequate version of PP recommends all and only those actions that can be consistently recommended by any other fully adequate version of PP. In other words, there can be no conflicting recommendations among fully adequate versions of PP. Or to put it yet another way, there is a unique consistent recommendation set associated with all fully adequate versions of PP. The following two lemmas facilitate this uniqueness result:

**Lemma 1**: For any S and H, $V = \langle S, H, A_\Omega\rangle$ is complete.

*Proof*: This is immediate from definition 8.

**Lemma 2**: Let $V = \langle S, H, A\rangle$ and $V^* = \langle S^*, H, A\rangle$ be versions of PP. If knowledge conditions are comparable and there is some action $a \in A_\Omega$ such that $a \in R(V^*)$ but $a \notin R(V)$, then $S^*$ is more stringent than S (i.e., $S^* \subset S$) and $R(V) \subset R(V^*)$.

*Proof*: Suppose that $V = \langle S, H, A\rangle$ and $V^* = \langle S^*, H, A\rangle$ are versions of PP, knowledge conditions are comparable and there is some action $a \in A_\Omega$ such that $a \in R(V^*)$ but $a \notin R(V)$. Then we need to show $S^* \subset S$ and $R(V) \subset R(V^*)$.

$\underline{S^* \subset S}$: From comparability of knowledge conditions, either $S = S^*$, $S \subset S^*$, or $S^* \subset S$. From our hypothesis, the first two of these three options result in a contradiction, leaving the third as the only possibility. If $S = S^*$, then $V$ and $V^*$ are identical, in which case, $R(V) = R(V^*)$, thereby contradicting the hypothesis that $a \in R(V^*)$ but $a \notin R(V)$. Suppose then that $S \subset S^*$. Since $a \notin R(V)$ it is either the case that $a \notin A$ or that there is a scenario $s$ in S whereby $a$ leads to an outcome in H. But a contradiction arises in either case. Since by hypothesis $a \in R(V^*)$, $a \in A$. And if $S \subset S^*$, the scenario $s$ in S whereby $a$ leads to an outcome in H is also in S*, which entails that $a \notin R(V^*)$, thereby also contradicting the hypothesis.

$R(V) \subset R(V^*)$: Since $a \in R(V^*)$ but $a \notin R(V)$, we only need to show that $R(V) \subseteq R(V^*)$. Consider any $a_i \in A_\Omega$ such that $a_i \in R(V)$. From definitions 4 and 5, $a_i \in A$ and there is no scenario in $S - S_d$ wherein $a_i$ leads to an outcome in H. Since A and H are the same in $V$ and $V^*$, to show $a_i \in R(V^*)$ it is only necessary to show that there is no scenario in $S^* - S^*_d$ wherein $a_i$ leads to an outcome in H. By definition 3, $s \in S^*_d$ if and only if $s \in S^*$ and, from every member of $A_\Omega$, $s$ leads to an outcome in H. Thus, since $S^* \subset S$, it follows from definition 3 that $S^*_d \subseteq S_d$, and therefore that $S^* - S^*_d \subseteq S - S_d$. Consequently, as there is no scenario in $S - S_d$ wherein $a_i$ leads to an outcome in H, there is similarly no such scenario in $S^* - S^*_d$, which gives us $R(V) \subseteq R(V^*)$. ■

> **Theorem 1 (Uniqueness)**: Let $V = \langle S, H_p, A\rangle$ and $V^* = \langle S^*, H_p, A^*\rangle$ be versions of PP, where $H_p$ is the harm condition representing the desired level of protection. If $V$ and $V^*$ are both fully adequate and knowledge conditions are comparable, then $R(V) = R(V^*)$.

*Proof*: By way of contradiction, suppose that the conditions of the theorem obtain but that $R(V) \neq R(V^*)$. Then there is an action $a \in A_\Omega$ such that $a \in R(V^*)$ but $a \notin R(V)$ or such that $a \in R(V)$ but $a \notin R(V^*)$. Without loss of generality, suppose that it is the former case, that is, $a \in R(V^*)$ but $a \notin R(V)$. Letting $V_\Omega = \langle S, H_p, A_\Omega\rangle$ and $V^*_\Omega = \langle S^*, H_p, A_\Omega\rangle$, the hypothesis that both $V$ and $V^*$ are fully adequate, and hence complete, yields $R(V_\Omega) = R(V)$ and $R(V^*_\Omega) = R(V^*)$. Thus, given the hypothesis $a \in R(V^*_\Omega)$ but $a \notin R(V_\Omega)$, lemma 2 entails that $R(V_\Omega) \subset R(V^*_\Omega)$. Because $R(V^*_\Omega) = R(V^*)$, this gives us $R(V_\Omega) \subset R(V^*)$. And since $R(V_\Omega) = R(V)$ and $V$ is non-paralyzing, $V_\Omega$ is also non-paralyzing. Moreover, by lemma 1, $V_\Omega$ is complete. Thus, there is a non-paralyzing and complete version of PP $V_\Omega = \langle S, H_p, A_\Omega\rangle$ such that $R(V_\Omega) \subset R(V^*)$. Consequently, $V^*$ does not satisfy condition iii of the definition of a fully adequate version of PP, which contradicts the hypothesis that $V^*$ is fully adequate. ■

That fully adequate versions of PP exist can be shown by example, as is illustrated in the next section.

## 2 Gardiner's Rawlsian maximin rule

As discussed in section 3.3.2, Gardiner (2006) states that the following conditions, inspired by Rawls's (1999) famous discussion of the original position, are sufficient to justify taking a precaution:

(a) No information exists upon which to judge the relative probability of members of S (pure uncertainty).

(b) There is no scenario in S in which catastrophe occurs when the precaution is taken.

(c) For every scenario in S, the precaution achieves results that are not significantly worse than the best that could have been achieved in that scenario.

(d) Every alternative to the precaution leads to catastrophe in at least one scenario in S.

Where S is a set of possible scenarios each of which satisfies some standard of scientific plausibility.

In section 3.3.2, I claimed that the interpretation of PP proposed here agrees with the maximin rule under these circumstances because the precaution is the only action that can be consistently recommended. That can be seen as follows. The relevant harm condition in this case is evidently catastrophe. Let $H_c$ denote the proper subset of $H_\Omega$ containing all and only those outcomes classified as catastrophic. Then it is easy to show that $V = \langle S, H_c, A_\Omega \rangle$ is a fully adequate version of PP and that the precaution is the sole member of $R(V)$. By condition (b), there is no scenario in S through which the precaution leads to an outcome in $H_c$, so $V$ does not recommend against the precaution by definition 4. Since the precaution is in $A_\Omega$, then, it follows from definition 5 that the precaution is in $R(V)$. By condition (d), for every alternative action to the precaution there is a scenario in S whereby it leads to an outcome in $H_c$. Hence, $V$ is obviously informative, as it recommends against every alternative action to the precaution.[3] From lemma 1, we can see that $V$ is complete. Finally, condition iii of the definition of a fully adequate version of PP is trivially satisfied in this case, because $R(V)$ is a singleton set. Consequently, any version of PP $V^*$ such that $R(V^*) \subset R(V)$ would be paralyzing. So $V$ is a fully adequate version of PP. And from the uniqueness theorem, we know that if the knowledge conditions are comparable, then for any other fully adequate version of PP $V^*$,

[3] Since there is no member of S wherein the precaution leads to an outcome in $H_c$, S contains no dismal scenarios and so $S_d = \varnothing$. By condition (d), then, for every alternative *a* to the precaution, there is a scenario in $S - S_d$ whereby *a* leads to an outcome in $H_c$, and consequently $V$ recommends against all of these alternatives.

$R(V^*) = R(V)$. In short, the precaution is the only thing that PP can consistently recommend given the Rawlsian conditions (a) through (d).

Section 3.3.2 noted that, given the interpretation of PP proposed here, the above reasoning suggests a more general sufficient condition for the precaution being the unique action that can be consistently recommended by PP. The more general sufficient condition is that an adequate version of PP consistently recommends the precaution, while no adequate version of PP consistently recommends any action other than the precaution. Translated into the formalism presented above, this means that there is an adequate $V$ such that the precaution is a member of $R(V)$, and no adequate $V^*$ such that $R(V^*)$ contains any action other than precaution. Since no adequate version of PP can be paralyzing, these two conditions entail that the consistent recommendation set of every adequate version of PP is a singleton containing only the precaution. In such cases, the precaution is the one and only thing that PP can recommend. However, there is no reason to suppose in general that the $R(V)$ of a fully adequate version of PP must contain only one member. When multiple actions can be consistently recommended by a fully adequate version of PP, efficiency is the guide for selecting among the remaining options.

## 3 Munthe's propositions 6 and 7 are equivalent

In his *The Price of Precaution and the Ethics of Risk* (2011, p. 92), Munthe states the following two propositions concerning responsible risk taking:

6. If, in a situation, there is a decision such that it will reduce the magnitude of a risk without significant costs compared to alternative decisions the agent could make, it is always less responsible (or, in a particular context of this sort, irresponsible) not to make that decision.
7. If a decision to introduce a risk is to be responsible, this decision must either produce some sufficiently substantial benefit or sufficiently reduce some risk.

I now show that 6 and 7 are logically equivalent. For convenience, if D is a decision to undertake a particular act (e.g., introducing a risk), then let ~D be the decision to abstain from that act. The proof, then, is as follows.

<u>6 ⇒ 7</u>: Let the decision D to introduce a risk be responsible. By way of contradiction, suppose that D neither produces a sufficiently substantial benefit nor sufficiently reduces any risk. But then ~D satisfies the antecedent of 6, as it avoids the risk introduced by D without significant costs compared to D. So, by 6, any decision incompatible with

~D is irresponsible. Hence, D is irresponsible, which contradicts our hypothesis. ■

$\underline{7 \Rightarrow 6}$: Consider a situation in which a decision D reduces the magnitude of a risk without significant costs compared to alternative decisions the agent could make. By way of contradiction, suppose that it is responsible to not choose D, i.e., that ~D is responsible. Then from proposition 7, ~D must either produce some sufficiently substantial benefit or sufficiently reduce some risk. But given the hypothesis, this cannot be the case, as D reduces risk without significant cost (i.e., loss of benefit or introduction of risk) compared to alternative decisions. ■

## 4 Triggering the precautionary principle

In section 4.4.2, I claimed that Munthe's proposition 7 is a straightforward consequence of the interpretation of PP proposed here, since an action that introduces a risk with no compensating benefit will be consistently recommended against by a version of PP. In this section, I use the formalization given above in section 1 to explicate this claim. The concept of consistently recommending against an action can be defined as follows:

9. A version of PP $V = \langle S, H, A \rangle$ *consistently recommends against* an action $a$ if and only if $V$ recommends against $a$ and $R(V) \neq \emptyset$.

In other words, $V$ recommends against $a$ while consistently recommending some other action. If $V$ consistently recommends against an action, then it is an *adequate* version of PP (i.e., neither paralyzing nor uninformative) but is not necessarily *fully adequate*.[4] A version of PP $V$ might consistently recommend against some action but fail to be fully adequate because it is not complete or because it uses a knowledge condition that is more stringent than necessary.

The conditions of Munthe's proposition 7 seem designed to ensure that the conditions of definition 9 are satisfied. The risk is said to be *introduced* and to produce no compensating benefit nor avoid any countervailing risk. Thus, the risk was avoidable, and some alternative action that abstains from imposing the risk, or any other equally serious risk, can be consistently recommended in its stead. Clearly, such an action would be consistently recommended against by some version of PP. Moreover, if some version of PP consistently recommends against an action, then every fully adequate version of PP utilizing the same harm condition does so too.

[4] Note, moreover, that any adequate version of PP must consistently recommend against some action.

**Theorem 2**: Let $V = \langle S, H_p, A\rangle$ be a version of PP, where $H_p$ is the harm condition representing the desired level of protection. If knowledge conditions are comparable and $V$ consistently recommends against $a \in A_\Omega$, then every fully adequate version of PP does so as well.

*Proof*: Let $V^* = \langle S^*, H_p, A^*\rangle$ be a fully adequate version of PP, and suppose that $V = \langle S, H_p, A\rangle$ consistently recommends against $a \in A_\Omega$. We want to show that $V^*$ also consistently recommends against $a$, that is, that $R(V^*) \neq \emptyset$ and $V^*$ recommends against $a$. Since $V^*$ is fully adequate by hypothesis, we know that $R(V^*) \neq \emptyset$, so only the second claim remains to be proven. By way of contradiction, suppose that $V^*$ does not recommend against $a$. Let $V_\Omega = \langle S, H_p, A_\Omega\rangle$, and $V^*_\Omega = \langle S^*, H_p, A_\Omega\rangle$. Since $V^*$ is fully adequate, and hence complete, $R(V^*_\Omega) = R(V^*)$. And since $A \subseteq A_\Omega$, $R(V) \subseteq R(V_\Omega)$. By hypothesis, $V$ recommends against $a$, which by definition 4 entails that $V_\Omega$ also recommends against $a$, and thus $a \notin R(V_\Omega)$. Because $V^*$ does not recommend against $a$ by our hypothesis for *reductio ad absurdum*, we have $a \in R(V^*)$ because $V^*$ is complete. And since $R(V^*_\Omega) = R(V^*)$, $a \in R(V^*_\Omega)$. Thus from lemma 2, we have that $R(V_\Omega) \subset R(V^*_\Omega)$. Since $R(V^*_\Omega) = R(V^*)$, this yields $R(V_\Omega) \subset R(V^*)$. Moreover, $V_\Omega$ is complete by lemma 1, and because $\emptyset \neq R(V) \subseteq R(V_\Omega)$, $V_\Omega$ is also non-paralyzing. Consequently, $V_\Omega$ is a non-paralyzing and complete version of PP such that $R(V_\Omega) \subset R(V^*)$. Hence, $V^*$ does not satisfy condition iii of the definition of a fully adequate version of PP, which contradicts the initial hypothesis that $V^*$ is fully adequate. ■

Theorem 2 tells us that once we have found a version of PP $V$ using the harm condition $H_p$ that consistently recommends against a course of action, we can be assured that no fully adequate version of PP could recommend in favor of that action. This is so even if $V$ is not fully adequate. That is helpful because it may be difficult to establish that a version of PP is complete and does not use a more-stringent-than-necessary knowledge condition. In such cases, theorem 2 assures us that actions consistently recommended against by $V$ are indeed excluded by PP. The concept of a version of PP consistently recommending against an action can also be used to render more precise talk of PP being "triggered" by some activity or proposed action:

10. An action *a triggers* PP if and only if there is a version of PP $V = \langle S, H_p, A\rangle$ that consistently recommends against $a$, where $H_p$ is the harm condition representing the desired level of protection.

Given theorem 2, an action that triggers PP is an action that PP recommends against. Notice that triggering PP may not be sufficient to indicate which action should be taken. If $V$ consistently recommends against some action, it may nevertheless be the case that $R(V)$ includes some actions that would be recommended against by fully adequate versions of PP. In addition, if $V$ is not complete, it may omit from its consistent recommendation set some actions that fully adequate versions of PP would include. And one of those overlooked actions might turn out to be the preferred option all things considered. Fully adequate versions of PP, then, delineate the full set of actions that are not consistently recommended against by a version of PP whose harm condition represents the desired level of protection.

## 5 Chichilnisky and sustainable welfare

The argument from excessive sacrifice, discussed in section 6.3.1, is a common defense of the pure time preference. In this section, I consider a sophisticated version of this argument due to Graciela Chichilnisky (1996), which argues that a pure time preference is necessary to avoid a "dictatorship of the future." Given that sustainability entails both preventing the interests of the present from overriding those of the future and vice versa, this proposal seems to provide a principled argument for the pure time preference. Here I explain why I reject Chichilnisky's account of sustainable welfare and the argument for the pure time preference embedded within it.

The basic element of Chichilnisky's proposal is a "utility stream," that is, a countably infinite sequence of real numbers each corresponding to the utility experienced by one generation. Thus, a utility stream might look like this: 0.4, 0.5, 0.3, 0.4, 0.6, 0.7, . . . . Welfare functions, denoted by $W(x)$, take utility streams as their arguments. Chichilnisky states two axioms for a sustainable welfare function: it may be neither a dictatorship of the present nor a dictatorship of the future (Chichilnisky 1996, pp. 240–1). Dictatorship of the present occurs if the welfare function is sensitive only to some finite segment of the utility stream (Chichilnisky 1996, p. 240). This axiom can be stated more precisely by means of the following notation. Let $\alpha_K$ be the initial segment of the utility stream $\alpha$ up to but not including K. Similarly, let $\gamma^K$ be the "tail" of the utility stream $\gamma$ from K on. Then $(\alpha^K, \gamma_K)$ is the utility stream that results from concatenating $\alpha_K$ and $\gamma^K$. The welfare function W, then, is a dictatorship of the present if the following condition holds: for any two utility streams $\alpha$ and $\beta$, $W(\alpha) > W(\beta)$ if and only if there is there is an N such that for every $K > N$, $W(\alpha^K, \gamma_K) > W(\beta^K, \sigma_K)$, for any utility streams $\gamma$ and $\sigma$. In other words, what happens *after* N makes no difference to the inequality

$W(\alpha) > W(\beta)$. Conversely, dictatorship of the future occurs if the welfare function is insensitive to the utilities of some initial finite segment of the infinite series of generations. More formally, for any two utility streams $\alpha$ and $\beta$, $W(\alpha) > W(\beta)$ if and only if there is there is an N such that for every K > N, $W(\gamma^K, \alpha_k) > W(\sigma^K, \beta_k)$, for any utility streams $\gamma$ and $\sigma$. In other words, what happens *before* N makes no difference to the inequality $W(\alpha) > W(\beta)$. A *sustainable welfare function*, then, is a complete sensitive preference ranking that is neither a dictatorship of the present nor a dictatorship of the future (Chichilnisky 1996, p. 241). A sensitive ranking is one with the following property: if two utility streams $\alpha$ and $\beta$ are identical except that the utility of one generation in $\alpha$ is higher than its counterpart in $\beta$, then $\alpha$ is preferred to $\beta$ (Chichilnisky 1996, p. 235). A sustainable welfare function, therefore, would avoid both the tyranny of the present and excessive demands of the future.

Chichilnisky shows that a number of previously proposed welfare functions are not sustainable according her definition. For example, a standard exponentially discounted function results in a dictatorship of the present, while some time-impartial welfare functions, such as the limit of the average utility, lead to a dictatorship of the future. In addition, Chichilnisky considers two welfare functions that could be regarded as possible interpretations of Parfit's suggestion that limits should be placed on how much sacrifice any generation can be required to make. These are a Rawlsian welfare function that prefers utility streams that maximize the utility of the least well-off generation, and a basic-needs welfare function requiring that the utility of each generation always attain some set minimum. Chichilnisky points out that these two welfare functions are not sustainable according to her definition: "Both . . . Rawlsian and basic needs criteria . . . are insensitive because they rank equally any two paths which have the same infimum even if one assigns much higher utility to the other generations" (Chichilnisky 1996, p. 245). In other words, the basic needs and Rawlsian welfare functions are rejected not on the grounds of being dictatorships of the future or present but because they are insensitive. We will see below that this argument depends crucially on the seeming harmless technical assumption that decisions are framed in reference to an infinite timescale.

Chichilnisky proves that sustainable welfare functions exist and that they must have the following form:

$$W(\alpha) = \sum_{g=1}^{\infty} \lambda_g \alpha_g + \phi(\alpha),$$

where $\forall g, \lambda_g > 0, \sum_{g=1}^{\infty} \lambda_g \alpha_g < \infty$, and $\phi(\alpha)$ is a purely finitely additive measure (Chichilnisky 1996, pp. 244–6). The term $\lambda_g$ represents the extent to which the utility of generation $g$ is discounted, and $\alpha_g$ denotes the utility of generation $g$. The condition $\sum_{g=1}^{\infty} \lambda_g \alpha_g < \infty$ entails that $\lambda_g$ goes to 0 as $g$ goes to infinity (i.e., the utility of future generations counts for less as time goes on). Thus, the first term on the right-hand side is a discounted welfare function, while the second term is a function that is impartial according to time, such as the limit of the average utility mentioned above. The central insight, then, is that every sustainable utility function must combine two elements: a future discount rate that enables it to avoid a dictatorship of the future and a time-impartial function that prevents a dictatorship of the present. Chichilnisky's argument, then, seems to provide a principled answer to Parfit's suggestion that the argument from excessive sacrifice could be adequately answered by a welfare function that respects intergenerational impartiality. Sustainability, according to Chichilnisky, requires a compromise between discounting and impartiality, a trade-off between the interests of the present and those of the future. And that in turn requires a pure time preference.

While I find Chichilnisky's proposal intriguing and ingenious, I also think that it rests on a fundamentally flawed assumption, namely that infinity is the relevant timescale for sustainability questions. To appreciate the implausibility of this assumption, note that the Earth will cease to exist in about 5 billion years and will be entirely uninhabitable by life of any sort in about 2.8 billion years (O'Malley-James *et al.* 2013). Whatever may be at stake when questions of environmental sustainability are raised, therefore, it cannot be the perpetuation of the Earth's natural resources for an infinite future. The implausibility of an infinite timescale is also brought out when one considers more closely what the terms "present" and "future" actually mean in Chichilnisky's analysis. The "present" is any initial finite segment, while the "future" is any "tail" of an infinite utility stream. For example, "present" could refer to the approximately 5-billion-year period from today's date until the time when the Sun becomes a red giant and incinerates the Earth. The "future," then, would be everything that happens in the infinite expanse of time after that. Yet this is obviously not what anyone means by "present" and "future" in discussions of sustainability or discounting. The present refers to *now*, or perhaps to a somewhat broader interval of time (e.g., 20 years) encompassing now, while discounting the future typically means discounting events 50, 100, 200, or 300 years ahead. Even astrobiologists refer to the end of the habitable stage of the Earth's lifespan as being in the "far future" (O'Malley-James *et al.* 2013)! Present and future, therefore, are normally understood in reference to a finite timescale. The future does not imply an infinite extension of time, but instead some reasonably large but finite number of years from now, where what counts as "large" may vary according to context.

Moreover, there is a strong epistemic reason for choosing a finite timescale in environmental policy and planning generally. Given our epistemic limitations, it is generally absurd to base policy judgments on expected impacts millions or billions of years into the future and beyond. The relevant timescale for policy decisions, therefore, must be restricted to some finite segment of time about which some inferences may be drawn concerning the potential consequences of our actions.

One might suppose that framing decisions in relation to an infinite future, while not literally accurate is nevertheless a harmless assumption that facilitates mathematical analysis. It is perhaps for this reason that the assumption of an infinite future has generally flown under the radar in discussions of Chichilnisky's formal theory of sustainability. However, I explain that the assumption of an infinite future as the relevant timescale is strongly substantive because it eliminates basic need and Rawlsian welfare functions as insensitive. To see this consider the following welfare function:

> *Basic needs plus average (BNPA)*: Criterion 1: Any utility stream in which each generation attains a specified basic level of subsistence is preferred over one in which that is not the case. Criterion 2: Any two streams not ordered by criterion 1 are ranked by their average utility (or limit of that average in the infinite case).

Observe that an analogous version of the Rawlsian maximin function could be constructed (i.e., utility streams are ordered by the maximin rule, and then average utility is used to break remaining ties). It is easily seen that BNPA is neither a dictatorship of the future nor of the present. Generations that fall below the basic needs threshold always matter to BNPA, no matter where in the utility stream they occur. But if an infinite future is the relevant timeframe for decisions, then BNPA is not *sensitive.* That is because in the infinite case, the utility of any single generation makes no difference to the limit of the average utility. Hence, in the infinite case, BNPA ranks equally any two utility streams in which all generations are above the basic needs threshold and are identical except that one generation is better off than its counterpart in the other stream. *But this insensitivity follows only if an infinite timescale is presumed.* If the timescale is finite, BNPA is obviously sensitive, since in that case the utility of each generation has some impact on the average. Thus, the argument against BNPA (and analogous Rawlsian welfare functions) as a sustainable welfare measure entirely rests on framing the decision problem in terms of an infinite future. Furthermore, basic need and Rawlsian welfare functions are compatible with intergenerational impartiality (as they do not require a pure discount rate) and straightforwardly express Parfit's idea that fairness places limits on how much sacrifice any generation can be required to make. Framing decisions in terms of an infinite future, therefore, cannot be viewed as a harmless technical assumption that simplifies the mathematics. Not only is such an assumption highly implausible for the reasons given above, it also unfairly stacks the deck against welfare functions, such as basic needs and a Rawlsian maximin, that respect intergenerational impartiality.[5] I conclude, therefore, that Chichilnisky's analysis of sustainability, despite its originality, does not constitute a good answer to Parfit's response to the argument from excessive sacrifice.

[5] Joerg Tremmel notes in relation to Rawls that theories of fairness or justice that respect intergenerational impartiality can function only in finite timescales (Tremmel 2009, pp. 158–70).

# *References*

Ackerman, F. 2008a. *Poisoned for Pennies: The Economics and Toxics of Precaution.* Washington, DC: Island Press.

Ackerman, F. 2008b. Hot, it's not: Reflections on *Cool It*, by Bjorn Lomborg. *Climactic Change* 89: 435–46.

Ackerman, F., and L. Heinzerling. 2004. *Priceless: On Knowing the Price of Everything and the Value of Nothing.* New York: The New Press.

Ackerman, F., and C. Munitz. 2012a. Climate damages in the FUND model: A disaggregated analysis. *Ecological Economics* 77: 219–24.

Ackerman, F., and C. Munitz. 2012b. Reply to Anthoff and Tol. *Ecological Economics* 81: 43.

Ackerman, F., and L. Stanton. 2012. Climate risks and carbon prices: Revising the social cost of carbon. *Economics: The Open-Access, Open-Assessment E-Journal* 6: article 10.

Ackerman, F., and L. Stanton. 2013. *Climate Economics: The State of the Art.* London: Routledge.

Ahteensuu, M., and P. Sandin. 2012. The precautionary principle. In *Handbook of Risk Theory*, Vol. 2, edited by S. Roeser, R. Hillerbrand, P. Sandin, and M. Peterson. Dordrecht: Springer, pp. 961–78.

Allen, M., J. Mitchell, and P. Stott. 2013. Test of a decadal climate forecast. *Nature Geoscience* 6: 243–44.

Anderson, E. 1993. *Value in Ethics and Economics.* Cambridge, MA: Harvard University Press.

Anderson, E. 2012. Epistemic justice as a virtue of social institutions. *Social Epistemology* 26(2): 163–73.

Anderson, M., S. Junankar, S. Scott, J. Jilkova, R. Salmons, and E. Christie. 2007. *Competiveness Effects of Environmental Tax Reform.* Brussels: European Commission.

Anthoff, D., and R. Tol. 2012. Climate damages in the FUND model: A comment. *Ecological Economics* 81: 42.

Arrow, K. 1999. "Discounting, Morality, and Gaming." In *Discounting and Intergenerational Equity*, edited by P. Portney and J. Weyant, 13–21. Washington, DC: Resources for the Future.

Arrow, K., and J. Hurwicz. 1972. "Decision making under ignorance." In *Uncertainty and Expectations in Economics: Essays in Honor of G.L.S. Shackle*, edited by C. Carter and J. Ford. Oxford: Basil Blackwell.

Aven, T. 2003. *Foundations of Risk Analysis: A Knowledge and Decision-Oriented Perspective*. Chichester, UK: John Wiley & Sons.

Aven, T. 2008. *Risk Analysis: Assessing Uncertainties beyond Expected Values and Probabilities*. Chichester, UK: John Wiley & Sons.

Aven, T. 2011. On different types of uncertainties in the context of the precautionary principle. *Risk Analysis* 31: 1515–25.

Baan, R., Y. Grosse, B. Lauby-Secretan, F. El Ghissassi, V. Bouvard, L. Benbrahim-Tallaa, N. Guha, F. Islami, L. Galichet, and K. Straif. 2011. Carcinogenicity of radiofrequency electromagnetic fields. *The Lancet Oncology* 12(7): 624–6.

Bantz, D. 1982. The philosophical basis of cost–risk–benefit analyses. In *PSA: The Proceedings of the Biennial Meeting of the Philosophy of Science Association*, edited by P. Asquith and T. Nickles. East Lansing, MI: Philosophy of Science Association, pp. 227–42.

Barham, B., J. Foltz, D. Jackson-Smith, and S. Moon. 2004. The dynamics of agricultural biotechnology adoption: Lessons from rBST use in Wisconsin, 1994–2001. *American Journal of Agricultural Economics* 86: 61–72.

Barrett, J. 2008. Approximate truth and descriptive nesting. *Erkenntnis* 68(2): 213–24.

Barrett, K., and C. Raffensperger. 1999. Precautionary science. In *Protecting Public Health and the Environment: Implementing the Precautionary Principle*, edited by C. Raffensperger and J. Tickner. Washington, DC: Island Press, pp. 106–22.

Beckerman, W., and C. Hepburn. 2007. Ethics of the discount rate in the Stern Review on the Economics of Climate Change. *World Economics* 8(1): 187–210.

Bedford, T., and R. Cooke. 2001. *Probabilistic Risk Assessment: Foundations and Methods*. Cambridge: Cambridge University Press.

Betz, G. 2013. In defence of the value free ideal. *European Journal for Philosophy of Science* 3: 207–20.

Biddle, J., and E. Winsberg. 2010. Value judgments in the estimation of uncertainty in climate modeling. In *New Waves in Philosophy of Science*, edited by P. Magnus and J. Busch. New York: Palgrave Macmillan, pp. 172–97.

Bigwood, E. 1973. The acceptable daily intake of food additives. *CRC Critical Reviews in Toxicology* June: 41–93.

Bigwood, E., and A. Gérard. 1968. *Fundamental Principles and Objectives of a Comparative Food Law: Vol. 2: Elements of Motivation and Elements of Qualification*. New York: S. Karger.

Billingsley, P. 1995. *Probability and Measure*, 3rd edn. New York: Wiley & Sons.

Bird, A., and J. Ladyman. 2012. *Arguing about Science (Arguing about Philosophy)*. New York: Routledge.

Black, H. 2005. Getting the lead out of electronics. *Environmental Health Perspectives* 113(10): A683–A685.

Blackburn, S., and K. Simmons, eds. 1999. *Truth*. Oxford: Oxford University Press.

Bognar, G. 2011. Can the maximin principle serve as a basis for climate change policy? *The Monist* 94: 329–48.

Braithwaite, R. B. 1953. *Scientific Explanation*. New York: Harper & Row.

Bratman, M. 1999. *Faces of Intention: Selected Essays on Intention and Agency*. Cambridge: Cambridge University Press.

Braverman, M. 2010. United States. In *The Use and Registration of Microbial Pesticides in Representative Jurisdictions Worldwide*, edited by J. Kabaluk, A. Svircev, M. Goettel, and S. Woo. IOBC Global, pp. 74–9.

Bridges, J., and O. Bridges. 2002. Hormones as growth promoters: The precautionary principle or a political risk assessment? In *The Precautionary Principle in the 20th Century: Late Lessons from Early Warnings*, edited by P. Harremoës *et al.* London: Earthscan Publications, pp. 161–9.

Broome, J. 1992. *Counting the Cost of Global Warming*. Cambridge: The White Horse Press.

Burnett, H. S. 2009. Understanding the precautionary principle and its threat to human welfare. *Social Philosophy and Policy* 26(2): 378–410.

Capper, J., E. Castañeda-Gutiérrez, R. Cady, and D. Bauman. 2008. The environmental impact of recombinant bovine somatotropin (rbST) use in dairy production. *Proceedings of the National Academy of Sciences* 105(28): 9668–73.

Case, D. 2005. The EPA's HPV challenge program: A tort liability trap? *Washington and Lee Law Review* 62: 147–205.

Chapman, P., A. Fairbrother, and D. Brown. 1998. A critical evaluation of safety (uncertainty) factors for ecological risk assessment. *Environmental Toxicology and Chemistry* 17: 99–108.

Chichilnisky, G. 1996. An axiomatic approach to sustainable development. *Social Choice and Welfare* 13(2): 231–57.

Chisholm, A., and H. Clarke. 1993. Natural resource management and the precautionary principle. In *Fair Principles for Sustainable Development: Essays on Environmental Policy and Developing Countries*, edited by E. Dommen. Cheltenham, UK: Edward Elgar, pp. 109–22.

Churchman, C. 1948. Statistics, pragmatics, induction. *Philosophy of Science* 15: 249–68.

Churchman, C. 1956. Science and decision making. *Philosophy of Science* 22: 247–9.

Clark, B., T. Phillips, and J. Coats. 2005. Environmental fate and effects of Bacillus thuringiensis (Bt) proteins from transgenic crops: A review. *Journal of Agricultural and Food Chemistry* 53: 4643–53.

Clarke, S. 2005. Future technologies, dystopic futures and the precautionary principle. *Ethics and Information Technology* 7: 121–6.

Cohen, J. 1992. *An Essay on Belief and Acceptance*. Oxford: Oxford University Press.

Cowen, T., and D. Parfit. 1992. Against the social discount rate. In *Justice between Age Groups and Generations*, edited by P. Laslett and J. Fishkin. New Haven: Yale University Press, pp. 144–61.

Cox, L. 2011. Clarifying types of uncertainty: When are models accurate, and uncertainties small? *Risk Analysis* 31: 1530–3.

Cranor, C. 1993. *Regulating Toxic Substances: A Philosophy of Science and the Law.* Oxford: Oxford University Press.

Cranor, C. 1999. Asymmetric information, the precautionary principle, and burdens of proof. In *Protecting Public Health and the Environment: Implementing the Precautionary Principle*, edited by C. Raffensberger and J. Tickner. Washington, DC: Island Press, pp. 74–99.

Cranor, C. 2001. Learning from law to address uncertainty in the precautionary principle. *Science and Engineering Ethics* 7: 313–26.

Cranor, C. 2004. Toward understanding aspects of the precautionary principle. *Journal of Medicine and Philosophy* 29(3): 259–79.

Cranor, C. 2006. *Toxic Torts: Science, Law and the Possibility of Justice.* Cambridge: Cambridge University Press.

Cranor, C. 2011. *Legally Poisoned: How the Law Puts Us at Risk from Toxicants.* Cambridge, MA: Harvard University Press.

Crump, K. 1984. A new method for determining allowable daily intakes. *Fundamental and Applied Toxicology* 4: 854–71.

Dana, D. 2009. The contextual rationality of the precautionary principle. *Northwestern University School of Law Public Law and Legal Theory Series,* No. 09–27.

Dasgupta, P., and G. Heal. 1979. *Economic Theory and Exhaustible Resources.* Cambridge: Cambridge University Press.

Diekmann, S., and M. Peterson. 2013. The role of non-epistemic values in engineering models. *Science and Engineering Ethics* 19: 207–18.

Dorato, M. 2004. Epistemic and nonepistemic values in science. In *Science, Values, and Objectivity*, edited by P. Machamer and G. Wolters. Pittsburgh, PA: University of Pittsburgh Press, pp. 52–77.

Dorne, J., and A. Renwick. 2005. The refinement of uncertainty/safety factors in risk assessment by the incorporation of data on toxicokinetic variability in humans. *Toxicological Sciences* 86: 20–6.

Dorne, J., K. Walton, and A. Renwick. 2001. Uncertainty factors for chemical risk assessment: Human variability in the pharmacokinetics of CYP1A2 probe substrates. *Food and Chemical Toxicology* 39: 681–96.

Douglas, H. 2000. Inductive risk and values in science. *Philosophy of Science* 67: 559–79.

Douglas, H. 2004. The irreducible complexity of objectivity. *Synthese* 138(3): 453–73.

Douglas, H. 2006. Bullshit at the interface of science and policy: Global warming, toxic substances, and other pesky problems. In *Bullshit and Philosophy*, ed. G. Hardcastle and G. Reisch. La Salle, IL: Open Court, pp. 213–26.

Douglas, H. 2007. Rejecting the ideal of value-free science. In *Value-Free Science?*, edited by H. Kincaid, J. Dupré, and A. Wylie. Oxford: Oxford University Press, pp. 120–39.

Douglas, H. 2008. The role of values in expert reasoning. *Public Affairs Quarterly* 22(1): 1–18.

Douglas, H. 2009. *Science and the Value-Free Ideal.* Pittsburgh, PA: University of Pittsburgh Press.

Dourson, M., S. Felter, and D. Robinson. 1996. The evolution of science-based uncertainty factors in noncancer risk assessment. *Regulatory Toxicology and Pharmacology* 24: 108–20.

Dourson, M., and J. Stara. 1983. Regulatory history and experimental support of uncertainty (safety) factors. *Regulatory Toxicology and Pharmacology* 3: 224–38.

Doyen, L., and J. Pereau. 2009. The precautionary principle as a robust cost-effectiveness problem. *Environmental Modeling and Assessment* 14: 127–33.

Earman, J. 1992. *Bayes or Bust? A Critical Examination of Bayesian Confirmation Theory.* Cambridge, MA: MIT Press.

Edqvist, L., and K. Pedersen. 2002. Antimicrobials as growth promoters: Resistance to common sense. In *The Precautionary Principle in the 20th Century: Late Lessons from Early Warnings*, edited by P. Harremoës *et al.* London: Earthscan Publications, pp. 100–10.

Elgie, S., and J. McClay. 2013. BC's carbon tax shift is working well after four years (attention Ottawa). *Canadian Public Policy* 39(s2): 1–10.

Elgin, C. 2004. True enough. *Philosophical Issues* 14: 113–21.

Elliott, K. 2010. Geoengineering and the precautionary principle. *International Journal of Applied Philosophy* 24(2): 237–53.

Elliott, K. 2011a. *Is a little pollution good for you?* New York: Oxford University Press.

Elliott, K. 2011b. Direct and indirect roles for values in science. *Philosophy of Science* 78: 303–24.

Elliott, K. 2013. Douglas on values: From indirect roles to multiple goals. *Studies in History and Philosophy of Science* 44: 375–83.

Elliott, K., and M. Dickson. 2011. Distinguishing risk and uncertainty in risk assessments of emerging technologies. In *Quantum Engagements: Social Reflections of Nanoscience and Emerging Technologies*, edited by T. Zülsdorf, C. Coenen, A. Ferrari, U. Fiedeler, C. Milburn, and M. Wienroth. Heidelberg: AKA Verlag, pp. 165–17.

Elliott, K., and D. McKaughan. 2014. Nonepistemic values and the multiple goals of science. *Philosophy of Science* 81(1): 1–21.

Engelhardt, T., and F. Jotterand. 2004. The precautionary principle: A dialectical reconsideration. *Journal of Medicine and Philosophy* 29: 301–12.

Environmental Protection Agency. 2003. National ambient air quality standards for ozone: Final response to remand; final rule. *Federal Register* 68(3): 614–44.

EU. 2000. *Communication from the Commission on the Precautionary Principle.* Brussels: Commission of the European Communities.

EU. 2006. *Regulation (EC) No 1907/2006 of the European Parliament and of the Council of 18 December 2006 concerning the Registration, Evaluation, Authorisation and Restriction of Chemicals (REACH).* Brussels: European Parliament.

EU. 2013. *General Report on REACH*. Brussels: European Commission.

Farman, J. 2002. Halocarbons, the ozone layer and the precautionary principle. In *The Precautionary Principle in the 20th Century: Late Lessons from Early Warnings*, edited by P. Harremoës *et al.* London: Earthscan Publications, pp. 79–89.

Feder, B. 1995. Dow Corning in bankruptcy over lawsuits. *The New York Times* May 16: D6.

Fildes, R., and N. Kourentzes. 2011. Validation and forecasting accuracy in models of climate change. *International Journal of Forecasting* 27: 968–95.

Filipsson, A., S. Sand, J. Nilsson, and K. Victorin. 2003. The benchmark dose method: Review of available models, and recommendations for application in health risk assessment. *Critical Review of Toxicology* 33: 505–42.

Fischer, E., J. Jones, and R. von Schomberg, eds. 2006. *Implementing the Precautionary Principle: Perspectives and Prospects*. Cheltenham, UK: Edward Elgar.

Foltz, J., and H. Chang. 2002. The adoption and profitability of rbST on Connecticut dairy farms. *American Journal of Agricultural Economics* 84(4): 1021–32.

Forster, M., and E. Sober. 1994. How to tell when simpler, more unified, or less *ad hoc* theories will provide more accurate predictions. *British Journal for the Philosophy of Science* 45: 1–35.

Foster, C. 2011. *Science and the Precautionary Principle in International Courts and Tribunals*. Cambridge: Cambridge University Press.

Frank, R. 2005. *Microeconomics and Behavior*, 6th edn. Boston, MA: McGraw-Hill.

Frigg, R., L. Smith, and D. Stainforth. 2013. The myopia of imperfect climate models: The case of UKCP09. *Philosophy of Science* 80: 886–97.

Frisch, M. 2013. Modeling climate policies: A critical look at integrated assessment models. *Philosophy and Technology* 26: 117–37.

Fu, G., Z. Liu, S. Charles, Z. Xu, and Z. Yao. 2013. A score-based method for assessing the performance of GCMs: A case study of Southeastern Australia. *Journal of Geophysical Research: Atmospheres* 118(10): 4154–67.

Gardiner, S. 2006. A core precautionary principle. *Journal of Political Philosophy* 14: 33–60.

Gardiner, S. 2010. Ethics and global climate change. In *Climate Ethics: Essential Readings*, edited by S. Gardiner, S. Caney, D. Jamieson, and H. Shue. New York: Oxford University Press, pp. 3–35.

Gardiner, S. 2011. *A Perfect Moral Storm: The Ethical Tragedy of Climate Change*. New York: Oxford University Press.

Gardiner, S., S. Caney, D. Jamieson, and H. Shue, eds. 2010. *Climate Ethics: Essential Readings*. New York: Oxford University Press.

Gareau, B. 2013. *From Precaution to Profit: Contemporary Challenges to Environmental Protection in the Montreal Protocol*. New Haven, CT: Yale University Press.

Gatehouse, A., N. Ferry, M. Edwards, and H. Bell. 2011. Insect-resistant biotech crops and their impacts on beneficial arthropods. *Transactions of the Royal Society B* 366: 1438–52.

Gee, D., and M. Greenberg. 2002. Asbestos: From "magic" to malevolent mineral. In *The Precautionary Principle in the 20th Century: Late Lessons from Early Warnings*, edited by P. Harremoës *et al.* London: Earthscan Publications, pp. 49–63.

Giere, R. 2003. A new program for philosophy of science? *Philosophy of Science* 70: 15–21.

Gilbertson, M. 2002. Early warnings of chemical contamination of the Great Lakes. In *The Precautionary Principle in the 20th Century: Late Lessons from Early Warnings*, edited by P. Harremoës *et al.* London: Earthscan Publications, pp. 138–47.

Godard, O. 2000. De la nature du principe de précaution. In *Le Principe de Précaution: Significations et Conséquences*, edited by E. Zaccai and J. Missa. Brussels: Editions de L'Université de Bruxelles, pp. 19–38.

Goklany, I. 2001. *The Precautionary Principle*. Washington, DC: Cato Institute.

Gómez-Barbero, M., J. Berbel, and E. Rodríguez-Cerezo. 2008. Bt corn in Spain – the performance of the EU's first GM crop. *Nature Biotechnology* 26(4): 384–6.

Graham, J. 2001. Decision-analytic refinements of the precautionary principle. *Journal of Risk Research* 4: 127–41.

Graham, J., and J. Wiener, eds. 1995. *Risk versus Risk: Trade-offs in Protecting Health and the Environment*. Cambridge, MA: Harvard University Press.

Graham, J., and J. Wiener. 2008a. The precautionary principle and risk–risk tradeoffs: A comment. *Journal of Risk Research* 11(4): 465–74.

Graham, J., and J. Wiener. 2008b. Empirical evidence for risk–risk tradeoffs: A rejoinder to Hansen and Tickner. *Journal Risk Research* 11: 485–90.

Goldman, A. 1999. *Knowledge in a Social World*. Oxford: Clarendon Press.

Haack, S. 1998. *Manifesto of a Passionate Moderate*. Chicago: University of Chicago Press.

Hanemann, W. 2008. *What Is the Economic Cost of Climate Change?* Berkeley, CA: UC Berkeley eScholarship.

Hansen, S., L. Carlsen, and J. Tickner. 2007. Chemicals regulation and precaution: Does REACH really incorporate the precautionary principle? *Environmental Science and Policy* 10: 395–404.

Hansen, J., M. Sato, P. Kharecha, D. Beerling, V. Masson-Delmotte, M. Pagani, M. Raymo, D. Royer, and J. Zachos. 2008. Target atmospheric CO2: Where should humanity aim? *Open Atmospheric Science Journal* 2: 217–31.

Hansen, S., and J. Tickner. 2008. Putting risk–risk tradeoffs in perspective: A response to Graham and Wiener. *Journal of Risk Research* 11(4): 475–83.

Hansen, S., M. von Krauss, and J. Tickner. 2008. The precautionary principle and risk–risk tradeoffs. *Journal of Risk Research* 11(4): 423–64.

Hansson, S. 1993. The false promises of risk analysis. *Ratio* 6: 16–26.

Hansson, S. 1994. *Decision Theory: A Brief Introduction*. Stockholm: KTH Royal Institute of Technology.

Hansson, S. 1996. Decision making under great uncertainty. *Philosophy of the Social Sciences* 26(3): 369–8.

Hansson, S. 1997. The limits of precaution. *Foundations of Science* 2: 293–306.

Hansson, S. 1999. Adjusting scientific practices to the precautionary principle. *Human and Ecological Risk Assessment* 5: 909–21.

Hansson, S. 2003. Ethical criteria of risk acceptance. *Erkenntnis* 59(3): 291–309.

Hansson, S. 2009. From the casino to the jungle: Dealing with uncertainty in technological risk management. *Synthese* 168: 423–32.

Hare, R. 1981. *Moral Thinking: Its Method, Levels, and Point.* Oxford: Oxford University Press.

Harremoës, P., D. Gee, M. MacGarvin, A. Stirling, J. Keys, B. Wynne, and S. Guedes Vaz. 2002. *The Precautionary Principle in the 20th Century: Late Lessons from Early Warnings*. London: Earthscan Publications.

Harris, J., and S. Holm. 2002. Extending human lifespan and the precautionary paradox. *Journal of Medicine and Philosophy* 3: 355–68.

Harsanyi, J. 1983. Bayesian decision theory, subjective and objective probabilities, and acceptance of empirical hypotheses. *Synthese* 57: 341–65.

Harsanyi, J. 1985. Acceptance of empirical statements: A Bayesian theory without cognitive utilities. *Theory and Decision* 18: 1–30.

Hartzell, L. 2009. Rethinking the precautionary principle and its role in climate change policy. PhD dissertation, Stanford University.

Hartzell-Nichols, L. 2012. Precaution and solar radiation management. *Ethics, Policy, and Environment* 15: 158–71.

Hartzell-Nichols, L. 2013. From "the" precautionary principle to precautionary principles. *Ethics, Policy and Environment* 16: 308–20.

Hauser, C., and H. Possingham. 2008. Experimental or precautionary? Adaptive management over a range of time horizons. *Journal of Applied Ecology* 45: 72–81.

Hausman, D. 1992. *The Inexact and Separate Science of Economics*. Cambridge: Cambridge University Press.

Heil, J. 1983. Believing what one ought. *Journal of Philosophy* 80(11): 752–65.

Hempel, C. 1965. Science and human values. In *Aspects of Scientific Explanation and other Essays in the Philosophy of Science*. New York: The Free Press, pp. 81–96.

Holling, C., ed. 1978. *Adaptive Environmental Assessment and Management*. Chichester, UK: John Wiley & Sons.

Hourdequin, M. 2007. Doing, allowing, and precaution. *Environmental Ethics* 29: 339–58.

Howard, D. 2006. Lost wanderers in the forest of knowledge: Some advice on how to think about the relation between discovery and justification. In *Revisiting Discovery and Justification: Historical and Philosophical Perspectives on the Context Distinction*, edited by J. Schickore and F. Steinle. Dordrecht: Springer, pp. 3–22.

Howson, C., and C. Urbach. 1993. *Scientific Reasoning: The Bayesian Approach*, 2nd edn. Chicago: Open Court.

Hsu, S. 2011. *The Case for a Carbon Tax: Getting Past our Hang-ups to Effective Climate Policy*. Washington, DC: Island Press.

Hunt, S. 2011. Theory status, inductive realism, and approximate truth: No miracles, no charades. *International Journal for the Philosophy of Science* 25(2): 159–78.

Ibarreta, D., and S. Swan. 2002. The DES story: Long term consequences of prenatal exposure. In *The Precautionary Principle in the 20th Century: Late Lessons from Early Warnings*, edited by P. Harremoës *et al.* London: Earthscan Publications, pp. 90–9.

Infante, P. 2002. Benzene: A historical perspective on the American and European occupational setting. In *The Precautionary Principle in the 20th Century: Late Lessons from Early Warnings*, edited by P. Harremoës *et al.* London: Earthscan Publications, pp. 35–48.

Intemann, K. 2004. Feminism, underdetermination, and values in science. *Philosophy of Science* 72(5): 1001–12.

Interagency Working Group. 2010. *Technical Support Document: Social Cost of Carbon for Regulatory Impact Analysis Under Executive Order 12866.* Washington, DC: US EPA.

Interagency Working Group. 2013. *Technical Support Document: Technical Update of the Social Cost of Carbon for Regulatory Impact Analysis Under Executive Order 12866.* Washington, DC: US EPA.

Jaeger, W. 2005. *Environmental Economics for Tree Huggers and Other Skeptics.* Washington, DC: Island Press.

Jeffrey, R. 1956. Valuation and acceptance of scientific hypotheses. *Philosophy of Science* 23: 237–46.

Jeffrey, R. 2004. *Subjective Probability (the Real Thing).* Cambridge: Cambridge University Press.

Jevrejeva, S., J. C. Moore, and A. Grinsted. 2010. How will sea level respond to changes in natural and anthropogenic forcings by 2100? *Geophysical Research Letters* 37: L077033.

John, S. 2007. How to take deontological concerns seriously in risk–cost–benefit analysis: A re-Interpretation of the precautionary principle. *Journal of Medical Ethics* 33: 221–4.

John, S. 2010. In defence of bad science and irrational policies: An alternative account of the precautionary principle. *Ethical Theory and Moral Practice* 13: 3–18.

Johnson, A. 2012. Avoiding environmental catastrophes: Varieties of principled precaution. *Ecology and Society* 17(3): 9.

Jordan, A., and T. O'Riordan. 1999. The precautionary principle in contemporary environmental policy and politics. In *Protecting Public Health and the Environment: Implementing the Precautionary Principle*, edited by C. Raffensberger and J. Tickner. Washington, DC: Island Press, pp. 15–35.

Kahneman, D., I. Ritov, and D. Schkade. 2006. Economic preferences or attitude expressions? An analysis of dollar responses to public issues. In *The Construction of Preference*, edited by S. Lichtenstein and P. Slovic. Cambridge: Cambridge University Press, pp. 565–93.

Kaplan, M. 1981. A Bayesian theory of rational acceptance. *Journal of Philosophy* 78: 305–30.

Kelly, K. 2007a. Ockham's razor, empirical complexity, and truth finding efficiency. *Theoretical Computer Science* 383: 270–89.

Kelly, K. 2007b. A new solution to the puzzle of simplicity. *Philosophy of Science* 74: 561–73.

Kimmel, C., and D. Gaylor. 1988. Issues in qualitative and quantitative risk analysis for developmental toxicology. *Risk Analysis* 8: 15–20.

Kincaid, H., J. Dupré, and A. Wylie, eds. 2007. *Value-Free Science?* Oxford: Oxford University Press.

Kitcher, P. 1990. The division of cognitive labor. *Journal of Philosophy* 87(1): 5–21.

Kitcher, P. 2001. *Science, Truth, and Democracy*. Oxford: Oxford University Press.

Kitcher, P. 2011. *Science in a Democratic Society*. Amherst, NY: Prometheus Books.

Knight, F. 1921. *Risk, Uncertainty and Profit*. New York: Houghton, Mifflin Company.

Koppe, J., and J. Keys. 2002. PCBs and the precautionary principle. In *The Precautionary Principle in the 20th Century: Late Lessons from Early Warnings*, edited by P. Harremoës *et al.* London: Earthscan Publications, pp. 64–78.

Kourany, J. 2010. *Philosophy of Science after Feminism*. Oxford: Oxford University Press.

Kriebel, D., J. Tickner, P. Epstein, J. Lemons, R. Levins, E. Loechler, M. Quinn, R. Rudel, T. Schettler, and M. Stoto. 2001. The precautionary principle in environmental science. *Environmental Health Perspectives* 109(9): 871–6.

Kuhn, T. 1977. Objectivity, value judgment, and theory choice. In *The Essential Tension*. Chicago: University of Chicago Press, pp. 320–39.

Kysar, D. 2010. *Regulating from Nowhere: Environmental Law and the Search for Objectivity*. New Haven, CT: Yale University Press.

Lacey, H. 1999. *Is Science Value-Free? Values and Scientific Understanding*. London: Routledge.

Lacey, H. 2004. Is there a significant distinction between cognitive and social values? In *Science, Values, and Objectivity*, edited by P. Machamer and G. Wolters. Pittsburgh, PA: University of Pittsburgh Press, pp. 24–51.

Lambert, B. 2002. Radiation: Early warnings, late effects. In *The Precautionary Principle in the 20th Century: Late Lessons from Early Warnings*, edited by P. Harremoës *et al.* London: Earthscan Publications, pp. 26–34.

Laudan, L. 2004. The epistemic, the cognitive, and the social. In *Science, Values and Objectivity*, edited by P. Machamer and G. Wolters. Pittsburgh, PA: University of Pittsburgh Press, pp. 14–23.

Lehman, A., and O. Fitzhugh. 1954. 100-fold margin of safety. *Quarterly Bulletin of the Association of Food and Drug Officials of the US* 18(1): 33–5.

Lemons, J., K. Shrader-Frechette, and C. Cranor. 1997. The precautionary principle: Scientific uncertainty and type I and type II errors. *Foundations of Science* 2: 207–36.

Lempert, R. 2002. A new decision science for complex systems. *Proceedings of the National Academy of Sciences*. 99, suppl. 3: 7309–13.

Lempert, R., D. Groves, S. Popper, and S. Bankes. 2006. A general, analytic method for generating robust strategies and narrative scenarios. *Management Science* 52: 514–28.

Lempert, R., S. Popper, and S. Bankes. 2003. *Shaping the Next One Hundred Years: New Methods for Quantitative, Long-term Policy Analysis*. Santa Monica, CA: RAND.

Leopold, A. 1966. *A Sand County Almanac*. New York: Oxford University Press.

Levi, I. 1960. Must the scientist make value judgments? *Journal of Philosophy* 57: 345–57.

Levi, I. 1962. On the seriousness of mistakes. *Philosophy of Science* 29: 47–65.

Levi, I. 1967. *Gambling with Truth*. London: Routledge & Kegan Paul.

Levi, I. 1980. *The Enterprise of Knowledge*. Cambridge, MA: MIT Press.

Lewis, S., J. Lynch, and A. Nikiforov. 1990. A new approach to deriving v community exposure guidelines from "no-observed-adverse-effect levels." *Regulatory Toxicology and Pharmacology* 11: 314–30.

Lichtenstein, S., and P. Slovic, eds. 2006. *The Construction of Preference*. Cambridge: Cambridge University Press.

Lipworth, L., R. Tarone, and J. McLaughlin. 2004. Breast implants and fibromyalgia: A review of the epidemiologic evidence. *Annals of Plastic Surgery* 52: 284–7.

Lohmann, R., H. Stapleton, and R. Hites. 2013. Science should guide TSCA reform. *Environmental Science and Technology* 47: 8995–6.

Lomborg, B. 2001. *The Skeptical Environmentalist: Measuring the Real State of the World*. New York: Cambridge University Press.

Lomborg, B. 2007. *Cool It: The Skeptical Environmentalist's Guide to Global Warming*. New York: Knopf.

Lomborg, B. 2010. *Smart Solutions to Climate Change: Comparing Costs and Benefits*. New York: Cambridge University Press.

Longino, H. 1990. *Science as Social Knowledge*. Princeton, NJ: Princeton University Press.

Longino, H. 1996. Cognitive and non-cognitive values in science: Rethinking the dichotomy. In *Feminism, Science, and the Philosophy of Science*, edited by L. Hankinson Nelson and J. Nelson. Dordrect: Kluwer, pp. 39–58.

Longino, H. 2002. *The Fate of Knowledge*. Princeton, NJ: Princeton University Press.

Loomes, G., and R. Sugden. 1982. Regret theory: An alternative theory of rational choice under uncertainty. *Economic Journal* 92: 805–24.

Losey, J., L. Rayor, and M. Carter. 1999. Transgenic pollen harms Monarch larvae. *Nature* 399: 214.

Lövei, G., D. Andow, and S. Arpaia. 2009. Transgenic insecticidal crops and natural enemies: A detailed review of laboratory studies. *Environmental Entomology* 38(2): 293–306.

Luce, R., and H. Raiffa. 1957. *Games and Decision: Introduction and Critical Survey*. New York: John Wiley & Sons.

Machamer, P., and H. Douglas. 1999. Cognitive and social values. *Science and Education* 8: 45–54.

Machamer, P., and L. Ozbeck. 2004. The social and the epistemic. In *Science, Values and Objectivity*, edited by P. Machamer and G. Wolters. Pittsburgh, PA: University of Pittsburgh Press, pp. 78–89.

Machamer, P., and G. Wolters, eds. 2004. *Science, Values and Objectivity*. Pittsburgh, PA: University of Pittsburgh Press.

Maher, P. 1993. *Betting on Theories*. Cambridge: Cambridge University Press.

Manson, N. 2002. Formulating the precautionary principle. *Environmental Ethics* 24: 263–74.

Marchant, G. 2001a. The precautionary principle: An "unprincipled" approach to biotechnology regulation. *Journal of Risk Research* 4: 143–57.

Marchant, G. 2001b. A regulatory precedent for hormesis. *Human and Experimental Toxicology* 20: 143–4.

Marchant, G., and Mossman, K. 2004. *Arbitrary and Capricious: The Precautionary Principle in the European Union courts*. Washington, DC: American Enterprise Institute.

Massey, R. 2009. *Toxic Use Reduction Act Program Assessment: Executive Summary*. Toxic Use Reduction Institute Methods and Policy Report #26.

Mayo, D. 1996. *Error and the Growth of Scientific Knowledge*. Chicago: University of Chicago Press.

Mayo, D., and A. Spanos, A. 2006. Severe testing as a basic voncept in a Neyman-Pearson philosophy of induction. *British Journal of Philosophy of Science* 57(2): 323–57.

Mayo, D., and A. Spanos. 2009. *Error and Inference: Recent Exchanges on Experimental Reasoning, Reliability, and the Objectivity and Rationality of Science*. Cambridge: Cambridge University Press.

McBride, W., S. Short, and H. El-Osta. 2004. The adoption and impact of bovine somatotropin on U.S. dairy farms. *Review of Agricultural Economics* 26(4): 472–88.

McGarrity, T., and W. Wagner. 2008. *Bending Science: How Special Interests Corrupt Public Health Research*. Cambridge, MA: Harvard University Press.

McGarvin, M. 2002. Fisheries: Taking stock. In *The Precautionary Principle in the 20th Century: Late Lessons from Early Warnings*, edited by P. Harremoës *et al.* London: Earthscan Publications, pp. 10–25.

McKinnon, C. 2009. Runaway climate change: A justice-based case for precautions. *Journal of Social Philosophy* 40: 187–203.

McMullin, E. 1982. Values in science. In *Proceedings of the Biennial Meeting of the Philosophy of Science Association, Volume 1*, edited by P. Asquith and D. Nickles. East Lansing, MI: Philosophy of Science Association, pp. 3–28.

Mearns, L. 2010. Quantification of uncertainties of future climate change: Challenges and applications. *Philosophy of Science* 77(5): 998–1011.

Michaels, D. 2008. *Doubt Is Their Product: How Industry's Assault on Science Threatens Your Health*. Oxford: Oxford University Press.

Mielke, H., and S. Zahran. 2012. The urban rise and fall of air lead (Pb) and the latent surge and retreat of societal violence. *Environment International* 43: 48–55.

Miller, H. 2010. Feds freeze out frost fix. *Forbes* (published online: http://www.forbes.com/2010/01/29/frost-agriculture-epa-science-opinions-contributors-henry-i-miller.html).

Mitchell, S. 2004. The prescribed and proscribed values in science policy. In *Science, Values, and Objectivity*, edited by P. Machamer and G. Wolters. Pittsburgh, PA: University of Pittsburgh Press, pp. 245–55.

Mitchell, S. 2009. *Unsimple Truths: Science, Complexity, and Policy.* Chicago: University of Chicago Press.

Morrall, J. 1986. A review of the record. *Regulation* 10: 25–34.

Munthe, C. 2011. *The Price of Precaution and the Ethics of Risk.* Dordrecht: Springer.

Nagel, E. 1961. *The Structure of Science: Problems in the Logic of Scientific Explanation.* New York: Harcourt, Brace & World.

Naranjo, S. 2009. Impacts of Bt crops on non-target invertebrates and insecticide use patterns. *CAB Reviews: Perspectives in Agriculture, Veterinary Science, Nutrition and Natural Resources* 4: 1–11.

Nevin, R. 2000. How lead exposure relates to temporal changes in IQ, violent crime, and unwed pregnancy. *Environmental Research* 83: 1–22.

Nevin, R. 2007. Understanding international crime trends: The legacy of preschool lead exposure. *Environmental Research* 104: 315–36.

Neyman, J. 1950. *A First Course in Probability and Statistics.* New York: Henry Holt.

Nielsen, G., and S. Øvrebø. 2008. Background, approaches and recent trends for setting health-based occupation exposure limits: A minireview. *Regulatory Toxicology and Pharmacology* 51: 253–69.

Nordhaus, W. 2007. Critical assumptions in the Stern Review on Climate Change. *Science* 317(13): 201–2.

Nordhaus, W. 2008. *A Question of Balance: Weighing the Options on Global Warming Policies.* New Haven, CT: Yale University Press.

Nordhaus, W. 2012. Why the global warming skeptics are wrong. *The New York Review of Books*, March 22 issue.

Nordhaus, W., and J. Boyer. 2000. *Warming the World: Economic Models of Global Warming.* Cambridge, MA: MIT Press.

Norton, B. 2005. *Sustainability: A Philosophy of Adaptive Ecosystem Management.* Chicago: University of Chicago Press.

O'Brien, M. 2000. *Making Better Environmental Decisions: An Alternative to Risk Assessment.* Cambridge, MA: MIT Press.

O'Malley-James, J., J. Greaves, J. Raven, and C. Cockell. 2013. Swansong biospheres: Refuges for life and novel microbial biospheres on terrestrial planets near the end of their habitable lifetimes. *International Journal of Astrobiology* 12(2): 99–112.

Oreskes, N., and E. Conway. 2010. *Merchants of Doubt: How a Handful of Scientists Obscured the Truth on Issues from Tobacco Smoke to Global Warming.* New York: Bloomsbury Press.

Oreskes, N., K. Shrader-Frechette, and K. Belitz. 1994. Verification, validation, and confirmation of numerical models in the earth sciences. *Science* 263(5147): 641–6.

Oreskes, N., D. Stainforth, and L. Smith. 2010. Adaptation to global warming: Do climate models tell us what we need to know? *Philosophy of Science* 77: 1012–28.

Parfit, D. 1984. *Reasons and Persons*. Oxford: Clarendon Press.

Parry, M. L., O. F. Canziani, J. P. Palutikof, P. J. van der Dinden, and C. E. Hansen, eds. 2007. *Climate Change 2007: Impacts, Adaptation, and Vulnerability: Contribution of Working Group II to the Fourth Assessment Report of the Intergovernmental Panel on Climate Change*. Cambridge: Cambridge University Press.

Pearce, D. 1998. *Economics and Environment: Essays on Ecological Economics and Sustainable Development*. Cheltenham, UK: Edward Elgar.

Pearce, D., G. Atkinson, and S. Mourato. 2006. *Cost–Benefit Analysis and the Environment: Recent Developments*. Paris: OECD Publishing.

Pennock, R. 2001. *Intelligent Design Creationism and Its Critics*. Cambridge, MA: MIT Press.

Percival, R. 2006. Who's afraid of the precautionary principle? *Pace Environmental Law Review* 23(1): 21–61.

Peterson, M. 2002. What is a *de minimis* risk? *Risk Management* 4: 47–55.

Peterson, M. 2006. The precautionary principle is incoherent. *Risk Analysis* 26(3): 595–601.

Peterson, M. 2007. Should the precautionary principle guide our actions or our beliefs? *Journal of Medical Ethics* 33: 5–10.

Peterson, M. 2009. *An Introduction to Decision Theory*. Cambridge: Cambridge University Press.

Pleasants, J., and K. Oberhauser. 2013. Milkweed loss in agricultural fields because of herbicide use: Effect on the Monarch butterfly population. *Insect Conservation and Discovery* 6(2): 135–44.

Popper, K. 1963. *Conjectures and Refutations*. New York: Routledge & Kegan Paul.

Popper, K. 1966. *The Open Society and Its Enemies, Volume II*. Princeton, NJ: Princeton University Press.

Popper, S., R. Lempert, and S. Bankes. 2005. Shaping the future. *Scientific American* 292(4): 66–71.

Posner, R. 2004. *Catastrophe: Risk and Response*. New York: Oxford University Press.

Powell, R. 2010. What's the harm? An evolutionary theoretical critique of the precautionary principle. *Kennedy Institute of Ethics Journal* 20(2): 181–206.

Price, C. 1993. *Time, Discounting and Value*. Oxford: Blackwell.

Price, G., P. Valdes, and B. Sellwood. 1997. Quantitative palaeoclimate GCM validation: Late Jurassic and mid-Cretaceous case studies. *Journal of the Geological Society* 154: 769–772.

Priest, G. 1998. What's so bad about contradictions? *Journal of Philosophy* 95(8): 410–26.

Proctor, R. 1991. *Value-Free Science? Purity and Power in Modern Knowledge*. Cambridge, MA: Harvard University Press.

Quijano, R. 2003. Elements of the precautionary principle. In *Precaution, Environmental Science, and Preventive Public Policy*, edited by J. Tickner. Washington, DC: Island Press, pp. 21–8.

Raffensperger, C., and J. Tickner, eds. 1999. *Protecting Public Health and the Environment: Implementing the Precautionary Principle*. Washington, DC: Island Press.

Rahmstorf, S., G. Foster, and A. Cazenave. 2012. Comparing climate projections to observations up to 2011. *Environmental Research Letters* 7(4): 1–5.

Ramsey, F. 1928. A mathematical theory of saving. *Economic Journal* 38(152): 543–59.

Rawls, J. 1999. *A Theory of Justice*, revised edition. Cambridge, MA: Harvard University Press.

Reiss, J., and P. Kitcher. 2009. Biomedical research, neglected diseases, and well-ordered science. *Theoria* 66: 263–82.

Renwick, A., and N. Lazarus. 1998. Human variability and noncancer risk assessment: An analysis of the default uncertainty factor. *Regulatory Pharmacology and Toxicology* 27: 3–20.

Resnik, D. 2003. Is the precautionary principle unscientific? *Studies in the History and Philosophy of Biological and Biomedical Sciences* 34: 329–44.

Resnik, D. 2004. The precautionary principle and medical decision making. *Journal of Medicine and Philosophy* 29(3): 281–99.

Resnik, M. 1987. *Choices: An Introduction to Decision Theory*. Minneapolis: University of Minnesota Press.

Revesz, R., and M. Livermore. 2008. *Retaking Rationality: How Cost Benefit Analysis Can Better Protect the Environment and Our Health*. New York: Oxford University Press.

Reyes, J. 2007. Environmental policy as social policy? The impact of childhood lead exposure on crime. *The B.E. Journal of Economic Analysis & Policy* 7(1): Article 51.

Rooney, P. 1992. On values in science: Is the epistemic/non-epistemic distinction useful? In *Proceedings of the 1992 Biennial Meeting of the Philosophy of Science Association, Volume 2*, edited by D. Hull, M. Forbes, and K. Okruhlik. East Lansing, MI: Philosophy of Science Association, pp. 13–22.

Rudén, C., and S. Hansen. 2010. Registration, evaluation, and authorization of chemicals (REACH) is but the first step – How far will it take us? Six further steps to improve the European chemicals legislation. *Environmental Health Perspectives* 118(1): 6–10.

Rudner, R. 1953. The scientist qua scientist makes value judgments. *Philosophy of Science* 20: 1–6.

Sachs, N. 2011. Rescuing the strong precautionary principle from its critics. *University of Illinois Law Review* 2011: 1285–338.

Sagoff, M. 2004. *Price, Principle, and the Environment*. New York: Cambridge University Press.

Sagoff, M. 2008. *The Economy of the Earth: Philosophy, Law, and the Environment*, 2nd edn. Cambridge: Cambridge University Press.

Salmon, W. 1966. *The Foundations of Scientific Inference*. Pittsburgh, PA: Pittsburgh University Press.

Sandin, P. 1999. Dimensions of the precautionary principle. *Human and Ecological Risk Assessment* 5(5): 889–907.

Sandin P. 2005. Naturalness and *de mininis* risk. *Environmental Ethics* 27(2): 191–200.

Sandin, P. 2006. A paradox out of context: Harris and Holm on the precautionary principle. *Cambridge Quarterly of Healthcare Ethics* 15(2): 175–83.

Sandin, P., M. Peterson, S. Hansson, and C. Rudén. 2002. Five charges against the precautionary principle. *Journal of Risk Research* 5(4): 287–99.

Santillo, D., P. Johnston, and W. Langston. 2002. TBT and antifoulants: A tale of ships, snails, and imposex. In *The Precautionary Principle in the 20th Century: Late Lessons from Early Warnings*, edited by P. Harremoës *et al.* London: Earthscan Publications, pp. 148–60.

Satterfield, T., R. Gregory, S. Klain, M. Roberts, and K. Chan. 2013. Culture, intangibles and metrics in environmental management. *Journal of Environmental Management* 117: 103–14.

Sears, M., R. Hellmich, D. Stanley-Horn, K. Oberhauser, J. Pleasants, H. Mattila, B. Siegfried, and G. Dively. 2001. Impact of Bt corn pollen on Monarch butterfly populations: A risk assessment. *Proceedings of the National Academy of Sciences* 98(21): 11937–42.

Semb, A. 2002. Sulphur dioxide: From protection of human lungs to remote lake restoration. In *The Precautionary Principle in the 20th Century: Late Lessons from Early Warnings*, edited by P. Harremoës *et al.* London: Earthscan Publications, pp. 111–20.

Scheffler, S. 1994. *The Rejection of Consequentialism: A Philosophical Investigation of the Considerations Underlying Rival Moral Conceptions*, revised edition. Oxford: Clarendon Press.

Schelling, T. 1997. The cost of combating global warming: Facing the tradeoffs. *Foreign Affairs* 76(6): 8–14.

Schleiter, K. 2010. Silicone breast implant litigation. *Virtual Mentor* 12: 389–94.

Schuur, E., and B. Abbott. 2011. High risk of permafrost thaw. *Nature* 480: 32–3.

Shrader-Frechette, K. 1991. *Risk and Rationality: Philosophical Foundations for Populist Reforms*. Berkeley: University of California Press.

Shrader-Frechette, K. 2011. *What Will Work: Fighting Climate Change with Renewable Energy, Not Nuclear Power*. New York: Oxford University Press.

Singer, P. 1972. Famine, affluence, and morality. *Philosophy and Public Affairs* 1: 229–43.

Singer, P. 2002. *One World: The Ethics of Globalization*. New Haven, CT: Yale University Press.

Solomon, M. 2001. *Social Empiricism*. Cambridge, MA: MIT Press.

Solomon, S., D. Qin, M. Manning, Z. Chen, M. Marquis, K. B. Averyt, M. Tignor, and H. L. Miller, eds. 2007. *Climate Change 2007: The Physical*

*Science Basis: Contribution of Working Group I to the Fourth Assessment Report of the Intergovernmental Panel on Climate Change*. Cambridge: Cambridge University Press.

Solow, R. 1974. The economics of resources or the resources of economics. *American Economic Association* 64(2): 1–14.

Soule, E. 2004. The precautionary principle and the regulation of US food and drug safety. *Journal of Medicine and Philosophy* 29: 333–50.

Sprenger, J. 2012. Environmental risk analysis: Robustness is essential for precaution. *Philosophy of Science* 79: 881–92.

Starmer, C. 1996. Explaining risky choices without assuming preferences. *Social Choice and Welfare* 13(2): 201–13.

Steel, D. 2008. *Across the Boundaries: Extrapolation in Biology and Social Science*. New York: Oxford University Press.

Steel, D. 2009. Testability and Ockham's razor: How formal and statistical learning theory converge in the new riddle of induction. *Journal of Philosophical Logic* 38: 471–89.

Steel, D. 2010. Epistemic values and the argument from inductive risk. *Philosophy of Science* 77: 14–34.

Steel, D. 2011. Extrapolation, uncertainty factors, and the precautionary principle. *Studies in History and Philosophy of Biological and Biomedical Sciences* 42: 356–64.

Steel, D. 2013a. The precautionary principle and the dilemma objection. *Ethics, Policy and Environment* 16: 318–37.

Steel, D. 2013b. Acceptance, values, and inductive risk. *Philosophy of Science* 80: 818–28.

Steel, D. 2013c. Precaution and proportionality: A reply to Turner. *Ethics, Policy and Environment* 16: 341–5.

Steel, D., and K. Whyte. 2012. Environmental justice, values, and scientific expertise. *Kennedy Institute of Ethics Journal* 22: 163–82.

Steele, K. 2006. The precautionary principle: A new approach to public decision making? *Law, Probability and Risk* 5: 19–31.

Steinemann, A. 2001. Improving alternatives for environmental impact assessment. *Environmental Impact Assessment Review* 21(1): 3–21.

Stern, N. 2007. *The Economics of Climate Change: The Stern Review*. Cambridge: Cambridge University Press.

Stern, N. 2008. The economics of climate change. *The American Economic Review* 98(2): 1–37.

Stern, N., and C. Taylor. 2007. Climate change: Risk, ethics, and the Stern Review. *Science* 317(13): 203–4.

Stirling, A., and D. Gee. 2003. Science, precaution, and practice. *Public Health Reports* 117(6): 521–533.

Stirzaker, D. 2003. *Elementary Probability*, 2nd edn. Cambridge: Cambridge University Press.

Strevens, M. 2003. The role of the priority rule in science. *Journal of Philosophy* 100: 55–79.

Summer, J., L. Bird, and H. Smith. 2009. *Carbon Taxes: A Review of Experience and Policy Design Considerations*. Technical Report NREL/TP-6A2–47312. Oak Ridge, TN: US Department of Energy.

Sunstein, C. 2001. *Risk and Reason: Safety Law and the Environment*. Cambridge: Cambridge University Press.

Sunstein, C. 2005. *Laws of Fear: Beyond the Precautionary Principle*. Cambridge: Cambridge University Press.

Taverne, D. 2005. *The March of Unreason: Science, Democracy and the New Fundamentalism*. Oxford: Oxford University Press.

Tengs, T., M. Adams, J. Pliskin, D. Gelb Safran, J. Siegel, M. Weinstein, and J. Graham. 1995. Five-hundred life-saving interventions and their cost effectiveness. *Risk Analysis* 15: 369–90.

Thalos, M. 2009. There is no core to precaution. *Review Journal of Political Philosophy* 7(2): 41–9.

Thalos, M. 2012. Precaution has its reasons. In *Topics in Contemporary Philosophy 9: The Environment, Philosophy, Science and Ethics*, edited by W. Kabasenche, M. O'Rourke, and M. Slater. Cambridge, MA: MIT Press, pp. 171–84.

Thompson, P. 2007. *Food Biotechnology in Ethical Perspective*. Dordrecht: Springer.

Tickner, J. 1999. A map toward precautionary decision making. In *Protecting Public Health and the Environment: Implementing the Precautionary Principle*, edited by C. Raffensperger and J. Tickner. Washington, DC: Island Press, pp. 162–186.

Tickner, J. 2002. Developing scientific and policy methods that support precautionary action in the face of uncertainty: The Institute of Medicine Committee on agent orange. *Public Health Reports* 117(6): 534–45.

Tickner, J., ed. 2003. *Precaution, Environmental Science, and Preventive Public Policy*. Washington, DC: Island Press.

Tickner, J., and K. Geiser. 2004. The precautionary principle stimulus for solutions- and alternatives-based environmental policy. *Environmental Impact Assessment Review* 24: 801–24.

Tickner, J., and T. Gouveia-Vigeant. 2005. The 1991 cholera epidemic in Peru: Not a case of precaution gone awry. *Risk Analysis* 25(3): 495–502.

Tickner, J., and D. Kriebel. 2006. The role of science and precaution in environmental and public health policy. In *Implementing the Precautionary Principle: Perspectives and Prospects*, edited by E. Fischer, J. Jones, and R. von Schomberg. Cheltenham, UK: Edward Elgar, pp. 42–62.

Tol, R. 2005. The marginal damage costs of carbon dioxide emissions: An assessment of the uncertainties. *Energy Policy* 33: 2064–74.

Tol, R. 2010. Carbon dioxide mitigation. In *Smart Solutions to Climate Change*, edited by B. Lomborg. Cambridge: Cambridge University Press, pp. 74–105.

Tremmel, J. 2009. *A Theory of Intergenerational Justice*. London: Earthscan.

Trouwborst, A. 2006. *Precautionary Rights and Duties of States*. Leiden: Martinus Nijhoff.

Tuana, N. 2010. Leading with ethics, aiming for policy: New opportunities for philosophy of science. *Synthese* 177: 471–92.

Turner, D. 2013. Proportionality and the precautionary principle. *Ethics, Policy and Environment* 16: 341–3.

Turner, D., and L. Hartzell. 2004. The lack of clarity in the precautionary principle. *Environmental Values* 13: 449–60.

UN. 1992. *Framework Convention on Climate Change*. New York: United Nations.

UNEP. 2010. *The Emissions Gap Report: Are the Copenhagen Pledges Sufficient to Limit Global Warming to 2°C or 1.5°C?* New York: United Nations.

US General Accounting Office. 2007. *Chemical Regulation: Comparison of U.S. and Recently Enacted European Union Approaches to Protect against the Risks of Toxic Chemicals*. GAO-07–825.

Valian, V. 1998. *Why So Slow? The Advancement of Women*. Cambridge, MA: MIT Press.

van Zwanenberg, P., and E. Millstone. 2002. Hormones as growth promoters: The precautionary principle or a political risk assessment? In *The Precautionary Principle in the 20th Century: Late Lessons from Early Warnings*, edited by P. Harremoës *et al.* London: Earthscan Publications, pp. 170–84.

vom Saal, F., and C. Hughes. 2005. An extensive new literature concerning low-dose effects of bisphenol A shows the need for a new risk assessment. *Environmental Health Perspectives* 113(8): 926–33.

von Krauss, M., and P. Harremoës. 2002. MTBE as a substitute for lead. In *The Precautionary Principle in the 20th Century: Late Lessons from Early Warnings*, edited by P. Harremoës *et al.* London: Earthscan Publications, pp. 121–37.

von Schomberg, R. 2006. The precautionary principle and its normative challenges. In *Implementing the Precautionary Principle: Perspectives and Prospects*, edited by E. Fischer, J. Jones, and R. von Schomberg. Cheltenham, UK: Edward Elgar, pp. 19–41.

Walters, C. 1986. *Adaptive Management of Renewable Resources*. New York: Macmillan.

Walton, K., J. Dorne, and A. Renwick. 2001a. Uncertainty factors for chemical risk assessment: Interspecies differences in in vivo pharmacokinetics and metabolism of human CYP1A2 substrates. *Food and Chemical Toxicology* 39: 667–80.

Walton, K., J. Dorne, and A. Renwik. 2001b. Uncertainty factors for chemical risk assessment: Interspecies differences in glucuronidation. *Food and Chemical Toxicology* 39: 1175–90.

Walton, K., J. Dorne, and A. Renwick. 2004. Species-specific uncertainty factors for compounds eliminated primarily through renal excretion in humans. *Food and Chemical Toxicology* 42: 261–74.

Weber, M. 1949. *On the Methodology of the Social Sciences*, translated by E. Shils and H. Finch. Glencoe, IL: The Free Press.

Weckert, J., and J. Moor. 2006. The precautionary principle in nanotechnology. *International Journal of Applied Philosophy* 2(2): 191–204.

Weitzman, M. 2001. Gamma discounting. *The American Economic Review* 91(1): 260–71.

Weitzman, M. 2007. A review of The Stern Review on the Economics of Climate Change. *Journal of Economic Literature* 45: 703–24.

Weitzman, M. 2009. On modeling and interpreting the economics of catastrophic climate change. *The Review of Economics and Statistics* 91(1): 1–19.

Werndl, C. 2009. What are the new implications of chaos for unpredictability? *British Journal for the Philosophy of Science* 60: 195–220.

Whiteside, K. 2006. *Precautionary Politics: Principle and Practice in Confronting Environmental Risk*. Cambridge, MA: MIT Press.

Williams, B. 2011. Passive and active adaptive management: Approaches and an example. *Journal of Environmental Management* 92: 1371–8.

Wright, J., K. Dietrich, M. Ris, R. Hornung, S. Wessel, B. Lanphear, M. Ho, and M. Rae. 2008. Association of prenatal and childhood blood lead concentrations with criminal arrests in early adulthood. *PLOS Medicine* 5(5): 732–9.

Zenghelis, D. 2010. Book review: *Smart Solutions to Climate Change: Comparing Costs and Benefits*. *Journal of Environmental Investing* 1(2): 79–89.

# Index


acid rain, 72
Ackerman, Frank, 53, 203–5
adaptive management, 63
agent-relative ethics, 142
agent-relative reason, 137–8
alternatives assessment, 66, 109–11
antimicrobials, 72
argument from inductive risk, 8, 11, 15, 144, 145, 146, 147–8, 150, 151, 152, 153, 154, 155, 157, 158, 159, 160, 161, 167, 170, 183, 185, 214, 215
  objections to, 149
Arrow, Kenneth, 53, 129–30
asbestos, 71, 73, 79
Aven, Terje, 99–100, 101–4

Bangladesh, 48
Bayesian decision theory, 107, 155
Beckerman, Wilfred, 133
benchmark dose (BMD), 174
Bendectin, 75, 77, 79
benzene, 71
Bigwood, E., 176
Bognar, Greg, 55
bovine growth hormone, 73, 200, 205–11
bovine spongiform encephalopathy (BSE), 73
British Columbia, 35
Broome, John, 122, 135
Bt (*Bacillus thuringiensis*) corn, 76, 78, 79
burden shifting reason, 138

carbon tax, 32–3, 34–5
Carson, Rachel, 72
Chichilnisky, Graciela, 233
Chisholm, Anthony, 58
Clarke, Harry, 58
climate change, 37, 71, 112, 113, 200–5
Cohen, Jonathan, 145–6, 149, 150–2, 153
condition, de minimis, 3, 37–8, 40, 220
consequentialism, 89–91
consistency, 15, 18, 28–9
cost-benefit analysis, 22, 25, 83, 96, 99, 108, 193, 205
  value commensurability, 113–18
  versus the precautionary principle, 85, 111
Cost-benefit analysis, 107
Cowen, Tyler, 120
Cox, Louis, 102–3
Cranor, Carl, 12, 82

Daubert ruling, 77
*de minimis* condition, 3
deontology, 4
dichlorodiphenyltrichloroethane (DDT), 72
Diekmann, Sven, and Martin Peterson, 169
diethylstilboestrol (DES), 72
discounting
  declining versus exponential, 140–1
  descriptive versus prescriptive approaches, 120–1, 142–3
  exponential, 123
    actual pure time preferences, 132
  pure time preference, 121–2, 124, 127, 128, 137, 139
    agent-relative ethics, 129–30, 133
    argument from excessive sacrifice, 128–9
    argument from special relations, 133
    argument that it is irrational, 131
    carbon tax, 141
    descriptive approach, 131
  Ramsey equation, 124–5, 126–7
  social discount rate, 123–4
discriminatory reason, 138
Douglas, Heather, 160, 171, 215
  direct versus indirect roles of values, 184–8
Dourson, Michael, and Jerry Stara, 176–7
Doyen, Luc, and Jean-Christophe Pereau, 63–5

efficiency, 18
electromagnetic fields (EMFs), 75, 77, 78, 79
Elliott, Kevin, 189, 191
environmental justice, 93, 211

Environmental Protection Agency (EPA), 72, 76, 87, 88, 89, 172, 174, 175, 193, 213
European Chemicals Agency (ECHA), 212
European Environment Agency (EEA), 69, 71

Food and Drug Administration (FDA), 72, 78, 169, 172, 205–6, 207, 211
Forster, Malcom, 168

Gardiner, Stephen, 7, 12, 45, 48, 53–6, 134, 140, 225
gasoline, leaded, 71
general circulation model (GCM), 104–5
Gérard, A., 176
Graham, John, 86–8

Hanemann, Michael, 204
Hansson, Sven Ove, 59, 61
Harris, John, and Søren Holm, 180
Hartzell-Nichols, Lauren, 5, 14, 31
  catastrophic precautionary principle (CPP), 49–52
Health Canada, 174, 206–7, 208, 210, 211
Hepburn, Cameron, 133
Hourdequin, Marion, 82
Hurwicz, Leonid, 53

"ice minus" bacteria, 76, 78, 79
Integrated Assessment Models (IAMs), 24
intelligent design creationism (IDC), 187
Interagency Working Group on the Social Cost of Carbon, 24, 203
intergenerational impartiality, 121–3, 127, 133, 134, 135, 139, 142
  versus discounting, 125–6
Intergovernmental Panel on Climate Change (IPCC), 31, 55, 105, 219

Jeffrey, Richard, 146, 158–60
Johnson, Alan, 66
Jordan, Andrew, 44

Kahneman, Daniel, 116
Kant, Immanuel, 94
Kitcher, Philip, 69, 184
Kriebel, David, 193–4
Kuhn, Thomas, 188

Lacey, Hugh, 146, 153, 157
Laudan, Larry, 167
Lehman, A., and O. Fitzhugh, 175–6
Leopold, Aldo, 66
Levi, Isaac, 153, 154, 156–7
Lomborg, Bjørn, 23, 35
Longino, Helen, 164, 167
Loomes, Graham, 56

Maher, Patrick, 154, 155
Manson, N., 33, 37
Marchant, Gary, 1, 44, 74–81
Massachusetts Toxics Use Reduction Act (TURA), 110
maximin rule, 2, 5, 7, 21, 25, 26, 53–6, 58–60, 66, 107, 225, 232
McKaughan, Dan, 191
McKinnon, Katherine, 12
measles, mumps, and rubella (MMR) vaccine, 28, 76, 77, 79
Merrell Dow Pharmaceuticals, 75
meta-precautionary principle (MPP), 83–4
methyl tert-butyl ether (MTBE), 72, 79
Mill, John Stuart, 94
minimax regret, 5, 56–62
Mitchell, Sandra, 67
monarch butterfly, 76
Montreal Protocol, 72, 136
Mossman, Kenneth, 1, 44, 74–81
Munthe, Christian, 38, 70, 89–91, 92, 93–4, 109, 226, 227

Neyman, Jerzy, 150
no-observed-adverse-effect level (NOAEL), 172
Nordhaus, William, 23, 24, 124, 205
Norton, Bryan, 63

O'Brien, Mary, 110–11
O'Riordan, Timothy, 44
ozone layer, 72
ozone, ground-level, 87–9

*P. syringae see* "ice minus" bacteria
Parfit, Derek, 120, 129, 133, 134–5, 142, 233
Peterson, Martin, 39, 41, 67
polybrominated diphenyl ethers (PBDEs), 71
polychlorinated biphenyls (PCBs), 71
Popper, Karl, 162
precautionary principle
  epistemic, 8, 160, 171, 172, 192–7, 214–17
  meta- (MPP), 9, 14
  objection, incoherence, 33
  procedural, decision, and epistemic versions, 10–11, 12
  tripod structure of, 10
  weak versus strong, 3, 17, 19–20
precautionary science, 7, 144, 170
preemptive war, 39
Principle 15 of the 1992 Rio Declaration on Environment and Development, 1, 67

proportionality, 4, 5, 6, 10, 11, 14, 18, 19, 26, 27, 33, 48, 50, 52, 63, 82, 84, 88–9, 176, 199
pure time preference, 121

Ramsey, F. P., 127–8
Rawls, John, 45, 53
recombinant bovine somatotropin (rBST) *see* bovine growth hormone
Registration, Evaluation, Authorization and Restriction of Chemicals (REACH), 200, 212–17
risk analysis, 96, 99, 108–13
risk trade-off analysis, 82, 86–8
risk versus uncertainty, 95–6, 108, 109, 194
Ritov, Ilana, 116
robust adaptive planning (RAP), 62–3, 65–8, 73
Rooney, Phyllis, 165
Rudner, Richard, 147–8, 153, 154

saccharin, 74, 79
safety factors *see* uncertainty factors
Sagoff, Mark, 115
Sandin, Per, 2, 3, 4, 9, 12, 18, 20, 37, 144, 145
Scheffler, Samuel, 130
Schkade, David, 116
scientific uncertainty, 9, 10, 11, 12, 14, 15, 17, 18, 21, 22, 23, 24, 25, 26, 47, 52, 62, 63, 65, 66, 83, 111, 182, 192, 199, 205, 211
  definition of, decision-theoretic, 97–101, 106, 107
  definition of, predictive validity, 96, 101–7
sequential justifiability, 121, 137
sequential plan, 121, 136–7
severity criterion, 170, 179, 181, 189
  applied to direct versus indirect role of values distinction, 185–7
silicone breast implants, 75, 79
Singer, Peter, 155
smog *see* ozone, ground-level
Sober, Elliott, 168
social cost of carbon (SCC), 22–3, 203
Solomon, Miriam, 165
Solow, Robert, 128
Sprenger, Jan, 67, 68
Stanton, Elizabeth, 203–5
Starmer, Chris, 115
Steele, Katie, 67
Stern, Nicholas, 22, 24, 124, 126, 128, 204, 205
Sugden, Robert, 56
sulfur dioxide, 72
Sunstein, Cass, 19, 25, 26, 33–4, 39, 116–17
sustainability, 229

Thalos, Mariam, 45, 46–8, 65, 109
Thompson, Paul, 207, 208, 211
Tickner, Joel, 193–4, 195
tin/lead solder, 110
Tol, Richard, 23, 24, 85, 205
Toxic Substances Control Act (TSCA), 212–13
tributyltin (TBT), 72
Trouwborst, Arie, 6, 109

uncertainty factors, 171, 172–5, 192–4
  empirical support, 175–8
United Nations Framework Convention on Climate Change (UNFCCC), 201
universalizability, 121, 122, 127, 129, 130, 142

values
  epistemic versus non-epistemic, 146, 161–5, 188
    objections, 165–8
  science-in-values standard
    objections, 192
    testability, 188
  value-free ideal, 7–8, 12, 15, 144–5, 146, 160, 167, 170, 171, 216
    risk assessment versus risk management, 197
  values-in-science standard, 171–2, 178–81
    applied to uncertainty factors, 182–4
    comparison with direct versus indirect roles of values, 185
von Schomberg, René, 194–7
*Vorsorgeprinzip*, 27

Wakefield, Andrew, 76
Weber, Max, 147
Weitzman, Martin, 128, 204
Wiener, Jonathan, 86–8
Wingspread Statement on the Precautionary Principle, 1, 69
Wittenoom, Australia, 73
World Health Organization, 174

Printed in the United States
By Bookmasters